Life Cycle Asses
Handb

Life Cycle Assessment Student Handbook

Edited by

Mary Ann Curran

WILEY

Copyright © 2015 by Scrivener Publishing LLC. All rights reserved.

Co-published by John Wiley & Sons, Inc. Hoboken, New Jersey, and Scrivener Publishing LLC, Salem, Massachusetts.
Published simultaneously in Canada.

No part of this publication may be reproduced, stored in a retrieval system, or transmitted in any form or by any means, electronic, mechanical, photocopying, recording, scanning, or otherwise, except as permitted under Section 107 or 108 of the 1976 United States Copyright Act, without either the prior written permission of the Publisher, or authorization through payment of the appropriate per-copy fee to the Copyright Clearance Center, Inc., 222 Rosewood Drive, Danvers, MA 01923, (978) 750-8400, fax (978) 750-4470, or on the web at www.copyright.com. Requests to the Publisher for permission should be addressed to the Permissions Department, John Wiley & Sons, Inc., 111 River Street, Hoboken, NJ 07030, (201) 748-6011, fax (201) 748-6008, or online at http://www.wiley.com/go/permission.

Limit of Liability/Disclaimer of Warranty: While the publisher and author have used their best efforts in preparing this book, they make no representations or warranties with respect to the accuracy or completeness of the contents of this book and specifically disclaim any implied warranties of merchantability or fitness for a particular purpose. No warranty may be created or extended by sales representatives or written sales materials. The advice and strategies contained herein may not be suitable for your situation. You should consult with a professional where appropriate. Neither the publisher nor author shall be liable for any loss of profit or any other commercial damages, including but not limited to special, incidental, consequential, or other damages.

For general information on our other products and services or for technical support, please contact our Customer Care Department within the United States at (800) 762-2974, outside the United States at (317) 572-3993 or fax (317) 572-4002.

Wiley also publishes its books in a variety of electronic formats. Some content that appears in print may not be available in electronic formats. For more information about Wiley products, visit our web site at www.wiley.com.

For more information about Scrivener products please visit www.scrivenerpublishing.com.

Cover design by Kris Hackerott

Library of Congress Cataloging-in-Publication Data:

ISBN 978-1-119-08354-2

Printed in the United States of America

10 9 8 7 6 5 4 3 2 1

Contents

Preface

This student handbook was created to serve as a companion to the 2012 *Life Cycle Assessment Handbook*[1], a compilation of writings by eminent leaders in the field of LCA and related methodology. The LCA Handbook was designed to be as comprehensive as possible, covering every facet of LCA methodology and presenting a variety of applications. This was quite a challenge given the ever-growing scope and acceptance of LCA over the years as an environmental management tool. The final product far exceeded my initial expectation. The chapter authors provided clear insight into the various aspects of LCA methodology and practice, and they openly shared their invaluable wisdom, experience and knowledge. However, the LCA Handbook does not attempt to explain in step-wise fashion how the various phases of an LCA can be completed. Other similar books and documents have also been published on LCA reflecting the ISO-standard[2] approach. But, again, few "how-to" guides exist. This student handbook is intended to fill that gap by addressing the individual steps of conducting, interpreting, and reporting an LCA.

For the sake of consistency, and maintaining a uniform "voice," the student handbook repeats much of the text prepared by the experts who contributed to the LCA Handbook. Because of the way in which the LCA Handbook was compiled, the chapters reproduce much of the same background introductory descriptions and the discussions on key issues scattered throughout the book. The student handbook brings these parts together in the appropriate sequence so that the chapters and sections present procedural guidance for conducting an LCA.

The student handbook then builds upon the various aspects of LCA practice with pertinent exercises for the reader to complete in order to help reinforce the messages within the sections. These exercises intend to help students gain a better understanding of the details involved in conducting an LCA by putting them in the position of both commissioner and practitioner of an assessment. In most cases, the exercises are thought problems, rather than ones requiring calculations or precise solutions. The aim is to encourage readers to look closer at certain methodological issues and check their understanding of them.

After presenting a brief overview (Chapter 1), the student handbook delves into the details of the stages that comprise LCA methodology: goal and scope definition (Chapter 2), life cycle inventory (Chapter 3), life cycle impact assessment (Chapter 4), normalization, grouping, and weighting (Chapter 5), and interpretation (Chapter 6). Chapter 7 addresses forward thinking applications of LCA in Life Cycle Sustainability Assessment (LCSA)

[1] Life Cycle Assessment Handbook: A Guide for Environmentally Sustainable Products (2012) MA Curran (ed) Scrivener-Wiley Publishing; ISBN 9778-1-118-09972-8; 640 pages.

[2] ISO 14040:2006 Environmental Management - Life Cycle Assessment - Principles and Framework, International Standard, International Organization for Standardization, Geneva, Switzerland.

including the role of modeling social impacts. The final chapter (8) provides additional resources readers might find useful.

The handbook aims to focus on LCA methodology and not extend into related, yet tangential, topics such as the life cycle of buildings, or life-cycle (eco)design. Also, the student handbook does not address the application of exergy analysis to LCA. There are many other textbooks that the reader can refer to that cover this topic in detail. As mentioned, Chapter 7 does address the topic of Life Cycle Sustainability Assessment (LCSA), including Social LCA, which the LCA Handbook also covers. Although not in detail, the chapter introduces the topic in order to give readers an idea of the future direction that is expected for LCA as its application moves toward meeting sustainability goals.

My sincere thanks go to the authors of the chapters in the LCA Handbook, which form the basis of the student handbook. Readers are encouraged to refer to the LCA Handbook as needed. Each chapter of the student handbook begins with page references to the LCA Handbook to make this easier for the reader.

1. *Life Cycle Assessment Handbook* Chapters and Authors:

1 Environmental Life Cycle Assessment: Background and Perspective
Gjalt Huppes and Mary Ann Curran

Part 1: Methodology and Current State of LCA Practice

2 An Overview of the Life Cycle Assessment Method – Past and Future
Reinout Heijungs and Jeroen B. Guinée

3 Life Cycle Inventory Modeling in Practice
Beverly Sauer

4 Life Cycle Impact Assessment
Manuele Margni and Mary Ann Curran

5 Sourcing Life Cycle Inventory Data
Mary Ann Curran

6 Software for Life Cycle Assessment
Andreas Ciroth

Part 2: LCA Applications

7 Modeling the Agri-Food Industry with Life Cycle Assessment
Bruno Notarnicola, Giuseppe Tassielli and Pietro A. Renzulli

8 Exergy Analysis and its Connection to Life Cycle Assessment
Marc A. Rosen, Ibrahim Dincer and Ahmet Ozbilen

9 Accounting for Ecosystem Goods and Services in Life Cycle Assessment and Process Design
Erin F. Landers, Robert A. Urban and Bhavik R. Bakshi

10 A Case Study of the Practice of Sustainable Supply Chain Management
Annie Weisbrod and Larry Loftus

11 Life Cycle Assessment and End of Life Materials Management
Keith A. Weitz

environmental impact caused by pollutant releases. Levels of emissions across the nation have stayed constant or declined; hundreds of primary and secondary wastewater treatment facilities have been built; land disposal of untreated hazardous waste has largely stopped; hundreds of hazardous waste sites have been identified and targeted for cleanup; and the use of many toxic substances has been banned. Together, these actions have had a positive effect on the nation's environmental quality and have set an example for other nations. However, despite the combined achievements of the federal government, States and industry in controlling waste emissions which have resulted in a healthier environment, the further improvement of the environment has slowed.

Worldwide, the advancement of environmental protection strategies moving from end-of-pipe to pollution prevention and beyond has been steady. This evolution can be summarized by the following chronology:

Evolution of Environmental Protection	
Chronology	**Strategy**
1970's to 1980's	End-of-Pipe Treatment
Mid 1980's	Waste Minimization/Reduction
Early 1990's	Pollution Prevention/Cleaner Production
Mid 1990's	ISO Certification/Life Cycle Assessment
2000 and Beyond	Sustainable Development/Life Cycle Sustainability Assessment

This evolution follows a pattern of ever-broadening scope when thinking about environmental management. In the 1980's, the term "waste minimization", or "waste reduction," was defined as "Measures or techniques that reduce the amount of wastes generated during industrial production processes; term is also applied to recycling and other efforts to reduce the amount of waste going into the waste stream." However, much of the focus remained on recycling and other end-of-life activities. In 1990, it was replaced by the term "pollution prevention" (or "cleaner production" outside the US) in order to give equal emphasis to activities that reduce potential environmental releases at the source of generation (Pollution Prevention Act 1990):

"The term "source reduction" means any practice which –

i. reduces the amount of any hazardous substance, pollutant, or contaminant entering any waste stream or otherwise released into the environment (including fugitive emissions) prior to recycling, treatment, or disposal; and
ii. reduces the hazards to public health and the environment associated with the release of such substances, pollutants, or contaminants. The term includes equipment or technology modifications, process or procedure modifications, reformulation or redesign of products, substitution of raw materials, and improvements in housekeeping, maintenance, training, or inventory control."

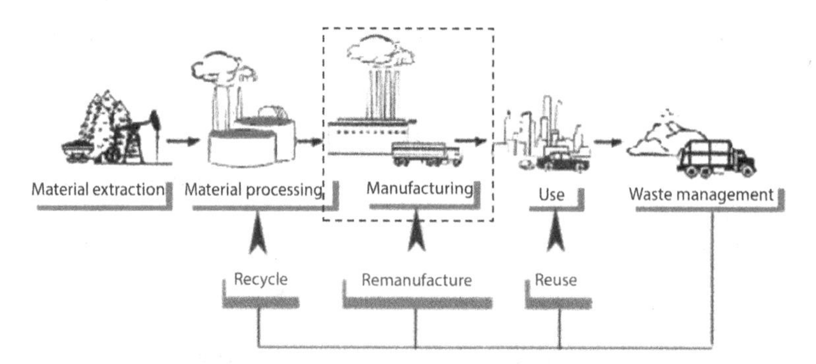

Figure 1.1 The boundaries of a pollution prevention (cleaner production) assessment are typically drawn around a single facility (dotted lines) omitting activities that may occur elsewhere in the product system.

However, the boundaries of a pollution prevention assessment[1] are drawn tightly around the facility or the plant (figure 1.1). This narrow, "gate-to-gate" focus does not allow for the identification of impacts that may occur in the manufacture and supply of materials going into the facility (i.e. the supply chain) or during the use and end-of-life stages of products coming out of the production facility.

Over the years, other federal policies have been developed to address environmental concerns at the various points across the life cycle. Some of these activities include the following:

- US National Environmental Policy Act (NEPA[2]) for Mining Operations.
- US EPA's March 1995 Risk Characterization Policy for assessing risk to human and ecological health from exposure to chemicals.
- The Resource Conservation and Recovery Act (RCRA), enacted in 1976, is the principal federal law in the United States governing the disposal of solid waste and hazardous waste.
- Energy Production and Use, for example, the Energy Star program that rates energy consuming products in order to help consumers optimize energy efficiency.
- Vehicles and Transportation, for example, the Renewable Fuels Standard which aims to replace conventional fossil fuels with those derived from bio-feedstock, such as bioethanol.

These examples demonstrate policy actions that focus on specific aspects. Like the fable about the six blind men and the elephant (Figure 1.2).

The conceptual jump to the broader environmental LCAs was made through a series of small steps. The first studies that are now recognized as (partial) LCAs date from the late 1960s and early 1970s, a period in which environmental issues like resource and energy efficiency, pollution control and solid waste became issues of broad public concern (US EPA 1993). One of the first (unfortunately unpublished) studies quantifying

---

[1] Pollution Prevention Assessment – Systematic, periodic internal reviews of specific processes and operations designed to identify and provide information about opportunities to reduce the use, production, and generation of toxic and hazardous materials and waste (US EPA 1992).

[2] Signed into law by President Richard Nixon on January 1, 1970.

Figure 1.2 The Six Blind Men and the Elephant.

the resource requirements, emission loadings and waste flows of different beverage containers was conducted by Midwest Research Institute (MRI) for the Coca Cola Company in 1969 (see box).

A follow-up of this study conducted by the same institute for the US EPA in 1974 (Hunt *et al* 1974), and a similar study conducted by Basler & Hofman (1974) in Switzerland, marked the beginning of the development of LCA as we know it today. MRI used the term Resource and Environmental Profile Analysis (REPA) for this kind of study, which was based on a system analysis of the production chain of the investigated products "from cradle to grave." After a period of diminishing public interest in LCA and a number of unpublished studies, there has been rapidly growing interest in the subject from the early 1980s on.

Lesson Learned from Aluminum Beverage Cans

In the early 1970s, the Coca-Cola Company conducted a study of its beverage containers. The results showed that all of the containers had some type of environmental impact. What Coca-Cola decided to do (from what was told to me) was not to ban or deselect the poorest-performing material(s). Instead, they challenged the material and container companies to make adjustments to their products and processes which would result in reduced life cycle environmental impacts over previous design options. For one of the materials – aluminum – the sector worked with local governments to develop a recycling infrastructure for the used beverage containers, resulting in a reduction of more than 90% in the energy used throughout the life cycle of the aluminum beverage container. The other material groups made similar improvements.

What did we learn? Because Coca-Cola chose not to ban any of the materials but challenged its suppliers instead, they created an innovative atmosphere which allowed development and financing of a recycling infrastructure to recapture the inherent value in the aluminum.

James Fava
PE International & Five Winds Strategic Consulting (now thinkstep)

The period 1970-1990 comprised the decades of conception of LCA with widely diverging approaches, terminologies and results. There was a clear lack of international scientific discussion and exchange platforms for LCA. During the 1970s and the 1980s LCAs were performed using different methods and without a common theoretical framework. LCA was repeatedly applied by firms to substantiate market claims. The obtained results differed greatly, even when the objects of the study were the same, which prevented LCA from becoming a more generally accepted and applied analytical tool (Guinée *et al* 2011).

The 1990s saw a remarkable growth of scientific and coordination activities world-wide, which is reflected in the number of workshops and other forums that have been organized in this decade and in the number LCA guides and handbooks produced:[3]

- *Product Life Assessments: Policy issues and implications*; Summary of a Forum on May 14, 1990; World Wildlife Fund and The Conservation Foundation: Washington, DC, 1990.
- Fava, J.A., Denison, R., Jones, B., Curran, M.A., Vigon, B., Selke, S., Barnum, J., Eds. *A Technical Framework for Life-Cycle Assessments*; Workshop Report Society of Environmental Toxicology and Chemistry; SETAC: Washington, DC, 1991.
- Smet, B. de, Ed. *Life-cycle analysis for packaging environmental assessment*; Proceedings of the specialised workshop, 24-25 September 1990, Leuven. Procter & Gamble Technical Center: Strombeek-Bever, Belgium, 1990.
- *Life-Cycle Assessment*; Proceedings of a SETAC-Europe workshop on Environmental Life Cycle Assessment of Products December 2-3 1991, Leiden; SETAC-Europe: Brussels, Belgium, 1992.
- Fava, J.A., Consoli, F., Denison, R., Dickson, K., Mohin, T., Vigon, B., Eds. *A Conceptual Framework for Life-Cycle Impact Assessment*; Society of Environmental Toxicology and Chemistry and SETAC Foundation for Environmental Education, Inc. Workshop Report; SETAC: Pensacola, Florida, 1993.
- Huppes, G., Schneider, F., Eds. Proceedings of the European Workshop on Allocation in LCA under the Auspices of SETAC-Europe, February 24-25, 1994, Leiden; SETAC-Europe: Brussels, Belgium, 1994.
- *Umweltprofile von Packstoffen und Packmitteln: Methode*; Fraunhofer-Institut für Lebensmitteltechnologie und Verackung: München; Gesellschaft für Verpackungsmarktforschung Wiesbaden und Institut für Energie- und Umweltforschung Heidelberg: Germany, 1991.
- Grieshammer, R., Schmincke, E., Fendler, R., Geiler, N., Lütge, E. Entwicklung eines Verfahrens zur ökologischen Beurteilung und zum Vergleich verschiedener Wasch- und Reinigungsmittel; Band 1 und 2. Umweltbundesamt: Berlin, Germany, 1991.
- *Product Life Cycle Assessment - Principles and Methodology*; Nord 1992:9, Nordic Council of Ministers: Copenhagen, Denmark, 1992.

---

[3] See also Chapter 8 Resources for Conducting Life Cycle Assessment

- Heijungs, R., Guinée, J.B., Huppes, G., Lankreijer, R.M., Udo de Haes, H.A., Wegener Sleeswijk, A., Ansems, A.M.M., Eggels, P.G., Duin, R. van, Goede, H.P. de. *Environmental life cycle assessment of products. Guide & Backgrounds - October 1992*; Centre of Environmental Science, Leiden University: Leiden, The Netherlands, 1992.
- Vigon, B.W., Tolle, D.A., Cornaby, B.W., Latham, H.C., Harrison, C.L., Boguski, T.L., Hunt, R.G., Sellers, J.D. *Life-Cycle Assessment: Inventory Guidelines and Principles*; EPA/600/R-92/245; Environmental Protection Agency: Washington, DC, 1993.
- Lindfors, L.-G., Christiansen, K., Hoffman, L., Virtanen, Y., Juntilla, V., Hanssen, O.J., Rønning, A., Ekvall, T., Finnveden, G. *Nordic Guidelines on Life-Cycle Assessment, Nord 1995:20*; Nordic Council of Ministers: Copenhagen, Denmark, 1995.
- Curran, M.A. *Environmental Life-Cycle Assessment*; McGraw-Hill: New York, 1996.
- Hauschild, M., Wenzel, H. Environmental Assessment of products. Volume 1: Methodology, tools and case studies in product development - Volume 2: Scientific background; Chapman & Hall: London, U.K., 1998.

Also, the first scientific journal papers started to appear in the *Journal of Cleaner Production, Resources, Conservation and Recycling, the International Journal of Life cycle Assessment, Environmental Science & Technology, the Journal of Industrial Ecology*, and other journals.

Through its North American and European branches, the Society of Environmental Toxicology and Chemistry (SETAC) started playing a leading and coordinating role in bringing LCA practitioners, users and scientists together to collaborate on the continuous improvement and harmonization of the LCA framework, terminology and methodology. The SETAC "Code of Practice" (Consoli *et al* 1993) was one of the key results of this coordination process. Next to SETAC, the International Standards Organization (ISO) has been involved in LCA since 1994. Whereas SETAC working groups focused at development and harmonization of methods, ISO adopted the formal task of standardizing methods and procedures.

The period of 1990-2000 can, therefore, be characterized as a period of *convergence* through SETAC's coordination and ISO's standardization activities, providing a standardized framework and terminology, and platform for debate and harmonization of LCA methods. In other words, the 1990s was a decade of standardization. Note, however, that ISO never aimed to standardize LCA methods in detail: "there is no single method for conducting LCA." During this period, LCA also became part of policy documents and legislation, with the main focus on packaging legislation, for example, in the European Union (EC 1994) and the 1995 Packaging Law in Japan (Hunkeler *et al* 1998).

1.4 Examples of Environmental Impact Trade-Offs

LCA identifies the potential transfer of environmental impacts from one medium to another (e.g., eliminating air emissions by creating a wastewater effluent instead) and/

or from one life cycle stage to another (e.g., from use and reuse of the product to the raw material acquisition stage). If an LCA were not performed, the transfer might not be recognized and properly included in the analysis because it is outside of the typical scope or focus of product selection processes. By broadening the study boundaries, LCA can help decision-makers select the product or process that causes the least impact to the environment. This information can be used with other factors, such as cost and performance data, in the selection process.

In connecting the different parts of the system, many LCAs lead to unexpected and non-intuitive results. For example, in the US in the 1980s, there was a perceived landfill crisis with many predicting the country running out of landfill space in the near future (NY Times 1986). Disposable (also called single-use) diapers (nappies) were caught up in the scare and perceived as a bad environmental choice because they end up in landfills by the millions, taking up valuable space, and take an estimated 500 years to decompose. Additionally, they are made using valuable non-renewable and renewable resources including wood pulp and plastic during their manufacture. But consumers often prefer the convenience and ease of disposable diapers.

Cloth diapers differ from disposables in that they are intended to be reused, thus cloth diapers are viewed as the more environmentally conscious alternative. While made of a renewable, natural material (cotton), cloth diapers require hot water (energy use) and detergents for washing. In order to determine the environmental superiority of cloth diapers, if any, multiple LCAs of disposable and cloth diapers were developed by P&G, the trade association EDANA, the UK Environment Agency, and others. However, when additional studies showed that cloth diapers also have meaningful environmental impacts due to use and heating water for washing, it became unclear which product was actually better.These studies found that most environmental impacts are linked to the energy, water, and detergents needed for cleaning cloth diapers, while the largest impacts, in addition to postconsumer waste, were related to raw material production for disposable diapers (Fava *et al* 1991, Krause *et al* 2009).

We learned that, depending upon the impact in question and where it occurs, different and equally valid interpretations can result. What these early studies revealed was that all products have impacts on the environment and that LCA tools enable decision makers to use new and additional information to make better-informed decisions.

Over the years, the instances in which one problem was solved but caused another are numerous. Compact fluorescent bulbs reduce electricity consumption by 75% but

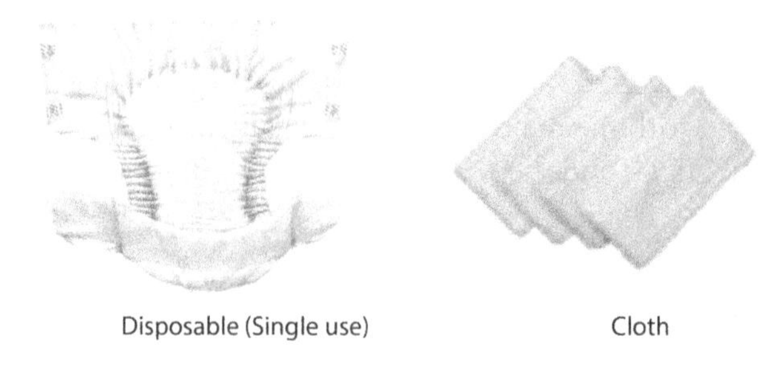

Figure 1.3 Dueling diaper (nappy) LCA studies raised awareness of the diversity of environmental impacts that products can create.

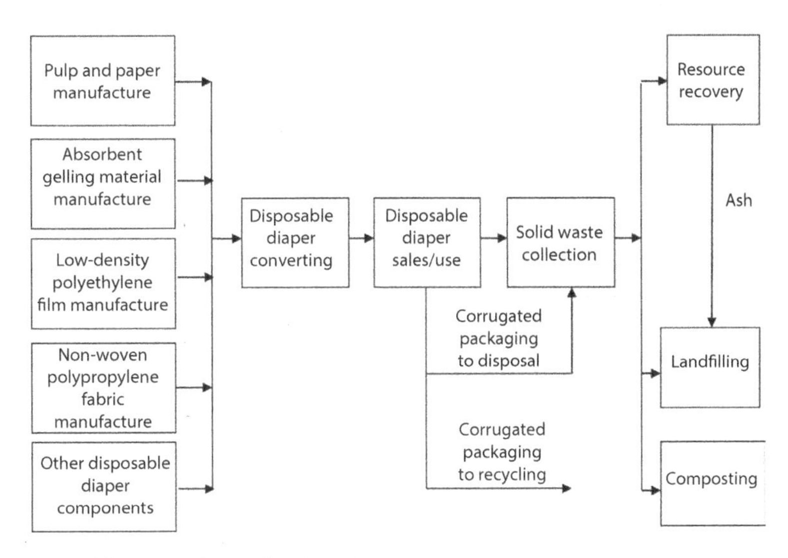

Figure 1.4 Disposable Diaper (Nappy) Life Cycle.

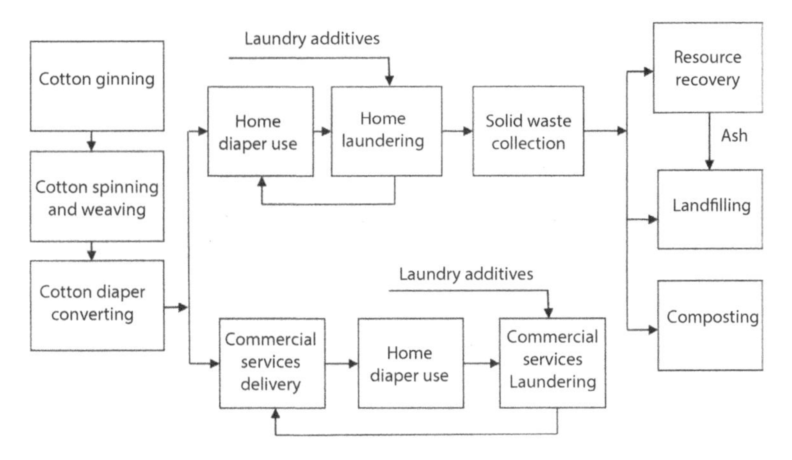

Figure 1.5 A Reusable Diaper (Nappy) Life Cycle.

come with a dash of mercury. Biobased fuels reduce greenhouse gas emissions but contribute to air, water and soil quality impacts in the agricultural stage.

Tools are needed that can help us to evaluate the comparative potential cradle-to-grave impacts of our actions in order to help us to prevent such wide-ranging effects. While LCA can provide assistance in the decision-making process, it has limited applicability in that it can only help us to evaluate the data that are available at the time. That is, it is not a predictive tool but can only model activities for which data are available. However, it has become increasingly evident that we must look much more holistically at our actions in order to more effectively protect human health and the environment in the short and long-term and to therefore, contribute to the development of more sustainable societies.

The Life Cycle of Methyl Tertiary-Butyl Ether (MTBE) as a Fuel Additive

MTBE is added to automotive fuel (gasoline/petrol) to increase octane levels and enhance combustion. It also provides the following environmental benefits:

- Reducing ozone precursors by 15%
- Reducing benzene emissions by 50%
- Reducing carbon monoxide emissions by 11%

But after it was commercialized, and the environmental benefits from reducing the emissions from vehicles were being realized, it became evident that there were measured amounts of MTBE in the environment. It could have leaked from storage tanks. MTBE in potable water supplies (e.g., lakes, reservoirs, and groundwater) is the greatest concern. Measured MTBE concentrations in some cases exceeded standard indicators for potable water, including "taste and odor" and "human health." There was insufficient information on its long-term toxicity.

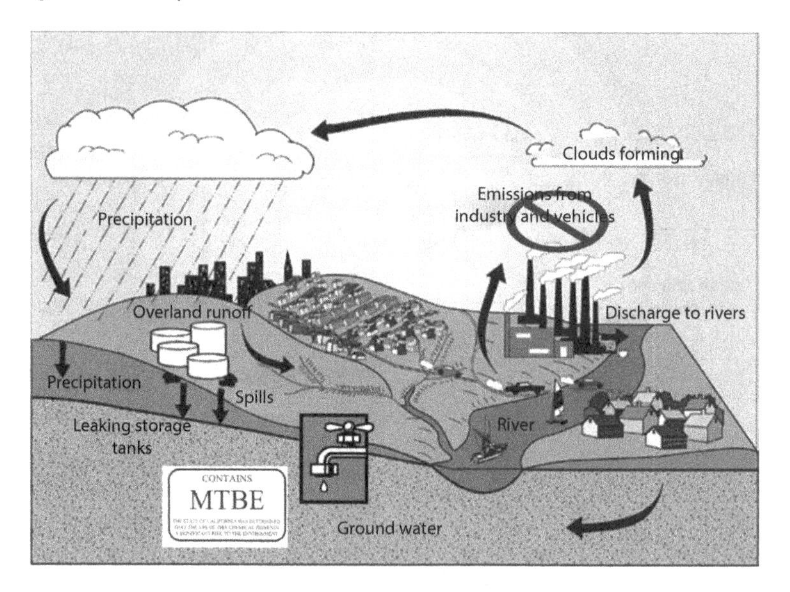

This graphic shows a system view of MTBE movement (modified from US Geologic Survey (http://sd.water.usgs.gov/nawqa/pubs/factsheet/fs114.95/fact.html)

While the use of MTBE reduced air emissions in the cities – an excellent outcome – it also created unexpected releases of MTBE into groundwater. Drinking water sources were contaminated as a result of an action that was designed to reduce air pollution.

As manufacturing operations become increasingly diverse, both technically and geographically, producers and the service industry are realizing the need to be fully aware of the potential environmental impacts in the sourcing of resources, manufacturing and assembly operations, usage, and final disposal. Many companies have found it advantageous to explore ways of moving *beyond* compliance using pollution prevention strategies and environmental management systems to improve their environmental

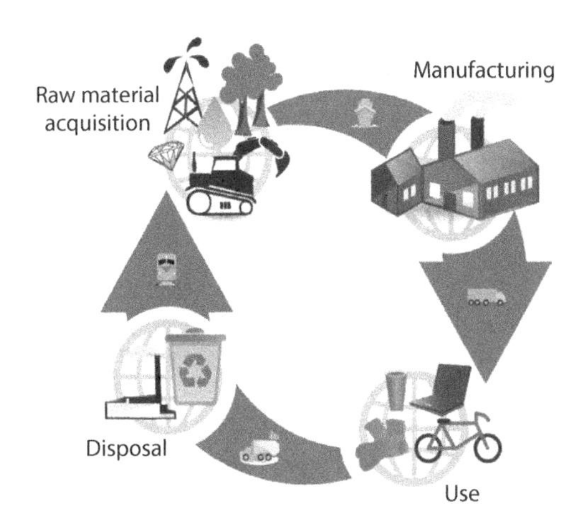

Figure 1.6 LCA is a "cradle-to-grave" assessment which spans the gathering of raw materials from the earth, manufacturing and use, on through to the return of materials to the earth. The arrows represent transportation.

performance. Society, in general, is becoming increasingly more aware of the fact that human activity can have far reaching impact.

This expanded view of interactions between human activity and the environment is prompting environmental managers and policy makers to look at products and services from cradle to grave. Out of this need came Life Cycle Assessment (LCA). What started as an approach to compare the environmental goodness (greenness) of products has developed into a standardized method for providing a sound scientific basis for environmental sustainability in industry and government. LCA provides a comprehensive view of the environmental aspects of product or process alteration or selection and presents an accurate picture of potential environmental trade-offs. LCA is useful in addressing cross-media problems and avoiding the transfer of a problem from one medium to another or from one place to another. Figure 1.6 presents a cradle- to-grave system of a generic product to depict the broad scope covered by LCA.

1.5 LCA Methodology

The LCA framework has evolved over time. In 1990, SETAC held the first in a series of LCA-related Pellston style workshops[4]. Although LCAs had been performed previously in one form or another, it was during this workshop when the name was coined and the resulting document presented the name of the method (SETAC 1990). As seen in Figure 1.7, the original LCA framework consisted only of three components with goal definition obviously missing. This omission was corrected in 1993 in a following SETAC

[4] Pellston workshops, named for the location of the first workshop of this type (Pellston, Michigan), aim to advance cutting edge technical and policy issues in environmental science by assembling scientists, engineers, and managers from government, private business, academia, and public interest groups to share current information on a given topic. At the end of the intense 4-5 day workshop, a document is produced that describes this knowledge with recommendations for enhancing the current state of the science.

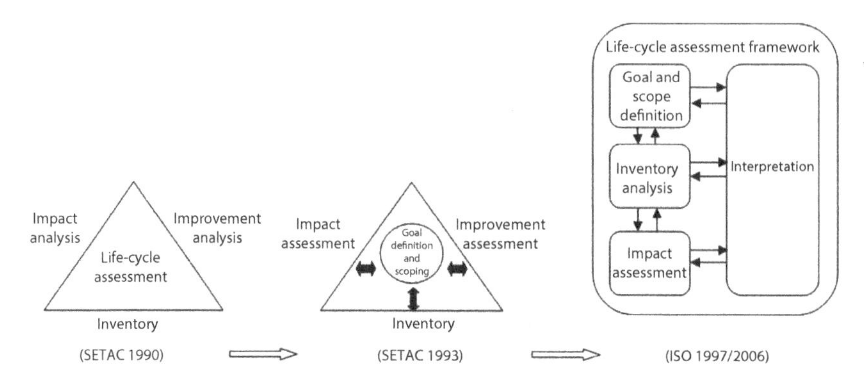

Figure 1.7 The Evolution of the Life Cycle Assessment Framework.

workshop, held in Sesimbra, Portugal. A new component called "Goal Definition and Scoping" was inserted in the middle of the SETAC triangle with arrows connecting it to Inventory, Impact Analysis, and Improvement Analysis, to depict the interconnections. By 1996, the triangle was replaced by a flow diagram with "Goal and Scope Definition" clearly shown as a first step (although the four interrelated phases of LCA are not necessarily conducted in 1, 2, 3, 4 order, GS&D should be addresses as a first step[5]).

The current LCA methodology refers to the process of compiling and evaluating the inputs, outputs and the potential environmental impacts of a product system throughout its life cycle. LCA has come a long way, and it continues to improve. Since a decade or so ago, there has been a broadly accepted set of principles that can be claimed as the present-day LCA framework.

The International Standards Organization (ISO) produced a series of standards and technical reports for LCA. Referred to as the 14040 series, these standards include the documents listed in Table 1.1.

The standards are organized into the different phases of an LCA study. These are:

> Goal and Scope Definition - identifying the purpose for conducting the LCA, the boundaries of the study, assumptions and expected output;
> Life Cycle Inventory - quantifying the energy use and raw material inputs and environmental releases associated with each stage of the life cycle;
> Life Cycle Impact Assessment - assessing the impacts on human health and the environment associated with the life cycle inventory results; and
> Interpretation – analysis of the results of the inventory and impact modelling, and presentation of conclusions and findings in a transparent manner.

The quality of a life-cycle inventory depends on an accurate description of the system to be analyzed. The necessary data collection and interpretation is contingent upon

[5] This was also when the component "Improvement Analysis" was renamed "Interpretation."

Table 1.1 ISO Documents on Life Cycle Assessment (LCA).

Number	Type	Title	Year
14040	International standard	Principles and framework	1996, 2006
14041	International standard	Goal and scope definition and inventory analysis	1998[1]
14042	International standard	Life cycle impact assessment	2000[1]
14043	International standard	Life cycle interpretations	2000[1]
14044	International standard	Requirements and guidelines	2006[2]
14047	Technical report	Examples of application of ISO 14042	2003
14048	Technical report	Data documentation format	2001
14049	Technical report	Examples of application of ISO 14041	2000

[1] Updated in 2006 and merged into 14044.

[2] Replaces 14041, 14042, and 14043.

Goal & Scope Definition:
- Determine the scope and system boundaries

Life Cycle Inventory:
- Data collection, modeling & analysis

Impact Assessment:
- Analysis of inputs/outputs using category indicators
- Group, normalize, weight results

Interpretation:
- Draw conclusions
- Checks for completeness, contribution, sensitivity analysis, consistency w/goal and scope, analysis, etc.

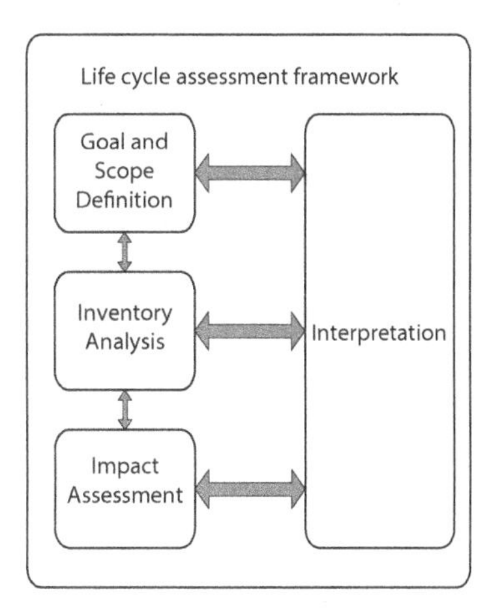

Figure 1.8 ISO Life Cycle Assessment Framework. This excerpt is adapted from ISO 14040:2006, Figure 1, page 8 with the permission of ANSI on behalf of ISO. © ISO 2015 - All rights reserved.

proper understanding of where each stage life-cycle begins and ends. The general scope of each stage can be described as follows:

Raw Materials Acquisition. This stage of the life cycle of a product includes the removal of raw materials and energy sources from the earth, such as the harvesting of trees or the extraction of crude oil. Land disturbance as well as transport of the raw materials from the point of acquisition to the point of raw materials processing are considered part of this stage.

Manufacturing. The manufacturing stage produces the product from the raw materials and delivers it to consumers. Three substages or steps are involved in this transformation: materials manufacture, product fabrication, and filling/packaging/distribution.

Materials Manufacture. This step involves converting raw material into a form that can be used to fabricate a finished product. For example, several manufacturing activities are required to produce a polyethylene resin from crude oil: The crude oil must be refined; ethylene must be produced in an olefins plant and then polymerized to produce polyethylene. Transportation between manufacturing activities and to the point of product fabrication should also be accounted for in the inventory, either as part of materials manufacture or separately.

Product Fabrication. This step involves processing the manufactured material to create a product ready to be filled, or packaged, for example, blow molding a bottle, forming an aluminum can, or producing a cloth diaper.

Filling/Packaging/Distribution. This step includes all manufacturing processes and transportation required to fill, package, and distribute a finished product. Energy and environmental wastes caused by transporting the product to retail outlets or to the consumer are accounted for in this step of a product's life cycle.

Use/Reuse/Maintenance. This is the stage consumers are most familiar with, the actual use, reuse, and maintenance of the product. Energy requirements and environmental wastes associated with product storage and consumption are included in this stage.

Recycle/Waste Management. Energy requirements and environmental wastes associated with product disposition are included in this stage, as well as postconsumer waste management options such as recycling, composting, and incineration.

The following general issues apply across all four life-cycle stages.

Energy and Transportation. Process and transportation energy requirements are determined for each stage of a product's life cycle. Some products are made from raw materials, such as crude oil, which are also used as sources for fuel. Use of these raw materials as inputs to products, represents a decision to forego their fuel value. The energy value of such raw materials that are incorporated into products typically is included as part of the energy requirements in an inventory. Energy required to acquire and process the fuels burned for process and transportation use is also included.

Environmental Waste Aspects. Three categories of environmental wastes are generated from each stage of a product's life cycle: atmospheric emissions, waterborne wastes, and solid wastes. These environmental wastes are generated by both the actual manufacturing processes and the use of fuels in transport vehicles or process operations.

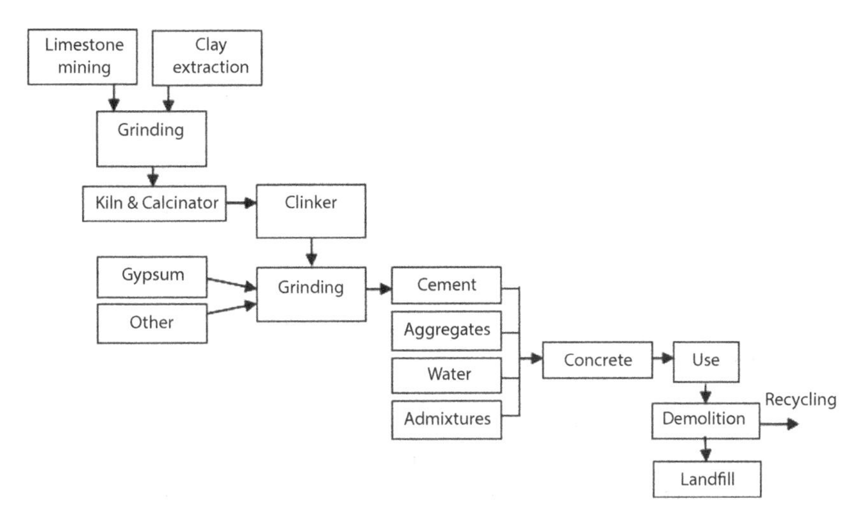

Figure 1.9 A Concrete Example of an LCA Flow Diagram (adapted from Sjunnesson 2005).

Waste Management Practices. Depending on the nature of the product, a variety of waste management alternatives may be considered: landfilling, incineration, recycling, and composting.

Allocation of Waste or Energy among Primary and Co-Products. Some processes in a product's life cycle may produce more than one product. In this event, energy and resources entering a particular process and all wastes resulting from it are allocated among the product and co-products. Allocation is described in more detail in Chapters 2 and 3.

1.6 Maintaining Transparency (Openness)

LCA involves various simplifying assumptions and value-based judgments throughout the process. LCAs can produce different results even if the same product seems to be the focus of the study. Differences can be caused by a number of factors, including:

- Different goal statements.
- Different functional units.
- Different boundaries.
- Different assumptions used to model the data.

The key is to keep these to a minimum and be explicit in the reporting phase about what assumptions and values were used. Readers of the study can then recognize the judgments and decide to accept, qualify, or reject them and the study as a whole.

It is very important to maintain transparency[6] in reporting an LCA study. The word transparent is used in the sense that it is easy to see what was done (versus the other

[6] ISO 14044 2006 defines transparency as the open, comprehensive and understandable presentation of information.

possible meaning of transparent which is to operate like a black box and be invisible to the user). This is necessary because it is not a single, prescriptive process. Rather, it involves multiple decision points that can greatly influence the outcome of the LCI and the LCIA. Although it would be best to achieve consensus on the methodology, thereby reducing or eliminating variations in the practice, at this time, the best solution is to maintain transparency and to fully document how the data were calculated. That way, even if others may not agree with the approach, at least, it is clear what was done.

1.7 Conclusions

If life cycle environmental information is to be integrated into product design and development to the same extent as price, quality, safety and performance, what are the changes that might need to occur? If we take a step back from these short case studies and critically examine what we have learned that might influence how we design and develop new products, technologies, and services for the 21st century, several breakthrough innovation principles and concepts surface are foundational to how we design and commercialize products.

1. Life cycle environmental performance will become as predominant as safety and quality are today in the design and development of products, technologies and services.
2. We have to create a marketplace which rewards greener companies, products, and brands.
3. We must move beyond examining a single impact or life cycle stage as the sole criteria for developing products that are more sustainable.
4. We already know much about life cycle hot spots and can direct innovation efforts by developing materials that incorporate the knowledge that exists in the life cycle community and in the thousands of LCA studies that have been completed over the last 20 years.
5. There is still significant additional information and knowledge to be learned, so we need to continue generating life cycle inventory data and conducting LCA studies. Moreover, there are questions that have not yet been formulated, let alone asked.

References

Basler & Hofman (1974) Studie Umwelt und Volkswirtschaft, Vergleich der Umweltbelastung von Behältern aus PVC, Glas, Blech und Karton; Basler & Hofman Ingenieure und Planer; Eidgenössisches Amt für Umweltschutz: Bern, Switzerland.

EC (1994) Directive 94/62/EC, OJ L 365, 31.12.1994, pp10-23; http://eur-lex.europa.eu/LexUriServ/LexUriServ.do?uri=CELEX:31994L0062:EN:HTML.

Consoli F, Alen D, Boustead I, Oude N de, Fava J, Franklin W, Quay B, Parrish R, Perriman R, Postlethwaite D, Seguin J,and Vigon B (eds) (1993) Guidelines for Life-Cycle Assessment: A Code of Practice, SETAC-Europe, Brussels, Belgium.

Fava, J.A., B. Denison, B. Jones, M.A. Curran, B. Vigon, S. Selke, J. Barnum (1991) *A Technical Framework for Life-Cycle Assessments.* SETAC Foundation for Environmental Education: Washington, D.C.

Guinée JB, Udo de Haes HA, Huppes G. (1993) Quantitative Life Cycle Assessment of Products 1: Goal Definition and Inventory. *J Clean Prod 1* (1):3-13.

Hunkeler D, Yasui I, Yamamoto R (1998) LCA in Japan: Policy and Progress. *Int J Life Cycle Assess* 3(3):124-130.

Hunt RG, Franklin WE, Welch RO, Cross JA, and Woodal AE (1974) *Resource and Environmental Profile Analysis of Nine Beverage Container Alternatives.* US Environmental Protection Agency: Washington, DC.

ISO (2006) ISO 14040 Environmental Management - Life Cycle Assessment - Principles and Framework, International Standard, International Organization for Standardization, Geneva, Switzerland.

Krause DR, Vachon S, and Klassen RD (2009) Special Topic Forum on Sustainable Supply Chain Management: Introduction and Reflections of the Role of Purchasing Management. *Journal of Supply Chain Management,* Vol 45:18–25.

The Life Cycle Assessment Handbook: A Guide for Environmentally Sustainable Products (2012) MA Curran (ed) Scrivener-Wiley; ISBN 9781118099728; 625 pp.

New York Times (1986) Garbage Crisis: Landfills are Nearly Out of Space, by Edward Hudson, Published: April 4, 1986

Pollution Prevention Act (1990) United States Code Title 42 The Public Health and Welfare Chapter 133.

US EPA (1992) Facility Pollution Prevention Guide, EPA/600/R-92-088, Office of Research and Development, Washington, DC.

US EPA (1993) Life Cycle Assessment: Inventory Guidelines and Principles EPA/600/R-92/245 Office of Research and Development, Washington, DC.

Why Take a Life Cycle Approach? (2004) Prepared for the UNEP/SETAC Life Cycle Initiative, United Nations Publications ISBN 92-807-24500-9; 24 pp.

Chapter 1 Exercises

1. Evolving Environmental Management
   Environmental management approaches has evolved over the years following a pattern of ever-broadening scope. Define and compare the scope and boundaries for the following:
   a) Pollutant Control Technology Assessment
   b) Waste Minimization (Cleaner Production) Assessment
   c) Greening the Supply Chain
   d) Product Life Cycle Assessment
   e) Life Cycle Sustainability Assessment
2. Seeing the Big Picture
   LCA is described as a valuable tool for identifying the potential transfer of environmental impacts, or burden shifting, resulting from making a change to a system or selecting one system over another. In the categories shown below, give examples of potential trade-offs that may occur. Use the sample scenarios provided or develop new ones.
   a) From one media to another (such as, air pollution control technology using filtration).
   b) From one life cycle stage to another (such as, postconsumer waste recycling).
   c) Between disparate environmental impacts (such as, comparing cloth and disposable diapers (nappies))
   d) Why are social impacts not included?
3. Modeling Assumptions
   In this chapter, cloth and disposable diapers (nappies) are used as an example of product comparison. While made of a renewable, natural material (cotton), cloth diapers require hot water (energy use) and detergents for washing. Disposable diapers contribute to postconsumer waste and are made from non-renewable raw materials.
   1. Describe how assumptions applied to the following modeling choices can affect the results:
      a) Product use (how many diapers (nappies) are used over what time period)?
      b) Location (where are the raw materials sourced and where is the product used)?
      c) Product performance (what other products should be considered for the two products to perform the same)?
      d) Other Considerations?
   2. Discuss how these modeling choices can affect the outcome of the study. What are the ethical considerations of deliberately introducing bias to arrive at a desired conclusion?

2

Goal and Scope Definition in Life Cycle Assessment

Abstract

Goal and scope definition is the starting point for any ISO-standardized Life Cycle Assessment (LCA). In this phase, the purpose of the assessment is established and decisions are made about the details of the product system being studied. These details are needed to guide the exact approach to be applied in conducting the assessment. The goal, as well as the scope, can be modified during the course of the work as data are collected and new information is revealed, e.g., it may be discovered that the proposed co-product allocation scheme does not work, not enough data are available to assemble a full life cycle inventory, etc. Such modifications should be (and in some cases, have to be) described transparently in the data spreadsheets and final report. The importance of this first phase according to ISO 14040/44 is often underestimated as it is much more than a simple introduction to the LCA process. Items to be defined in the goal and scope definition phase include the functional unit, system boundaries, granularity of the process data, exclusion of life cycle stages or inputs, and the selection of impact indicators and characterization factors.

This chapter describes the goal and scope definition phase of LCA including how to properly define the goal of an LCA, which then leads to defining the scope and boundaries of the system to be assessed. Examples of goal statements are provided.

References from the LCA Handbook

Aims of the Chapter

1. Describe how to properly define the intended goal of an LCA study.
2. Describe how to define the scope of an LCA study according to the goal.
3. Help the reader understand how the goal directly impacts data collection, impact modeling and interpretation efforts.

2.1 Introduction

A Life Cycle Assessment (LCA) starts with the definition of the intended goal and scope of the study, before proceeding to the inventory analysis, then optionally to impact assessment, and ending with interpretation. A successful outcome of any assessment depends on a clear, unambiguous definition of the purpose of the study from the outset. More than a simple introduction to a Life Cycle Assessment (LCA), goal and scope definition is an integral part of conducting an LCA and relates to any of the other phases. The structure of the LCA methodology has been well established by the International standard 14040 (2006). It clearly asserts the goal and scope definition phase as the first of four inter-related phases:

1. Goal and Scope Definition - Clearly defining the goal and scope of the study (including selecting a functional unit);
2. Inventory Analysis - Compiling an inventory of relevant energy and material inputs and environmental releases (Life Cycle Inventory (LCI) analysis);
3. Impact assessment - Evaluating the potential environmental impacts associated with identified inputs and releases (Life Cycle Impact Assessment (LCIA));
4. Interpretation - Interpreting the results to help decision makers make a more informed decision. Due to the iterative nature of LCA, it is important that in a result and interpretation phase of any of the relevant result aspects mentioned and conclusion aspects drawn must be already stated or mentioned in the goal and scope.

As already mentioned, but worth repeating, it is crucial for the goal of the study to be clearly defined at the outset. A well-defined goal will in turn help define the scope and boundaries of the study. The 2006 ISO standard states "the scope should be sufficiently well defined to ensure that the breadth, depth and detail of the study are compatible and sufficient to address the stated goal." It is also critical in directing future data collection efforts (the inventory analysis phase). The importance of properly defining the goal and scope of an LCA study cannot be overstated.

Conducting an LCA can help answer a number of important questions of concern to decision makers. Some examples include the following:

- What is the impact of a product system to particular interested parties and stakeholders?
- Which product or process causes the least environmental impact, overall or in each stage of its life cycle?
- How might changes to the current system affect the environmental impacts across all life cycle stages?
- Which technology or process causes the least amount of acid rain, smog formation, or damage to local trees (or any other impact category of concern)?
- How can the process be changed to reduce a specific environmental impact of concern (e.g., global warming)?

However, LCA is a highly iterative process, so that the practitioner may need to go back to goal and scope after the preliminary life cycle inventory (LCI) work, move back from impact assessment to inventory analysis, or have a look at the interpretation in the early stages, etc.

LCAs can be conducted to gather data and information describing a single system in order to establish a baseline of impacts, but most are performed to compare two or more systems. Study results may be intended for internal use or for sharing with external parties. Examples of types of LCAs and goals include the following:

- **Single System – Internal Use of Results**
  - Analyze current product to identify opportunities for reducing environmental impact
  - Establish product baseline against which to measure future improvements
- **Single System – External Use of Results**
  - Environmental product declaration (e.g., to share with customers who request information about environmental metrics for product)
- **Comparative Analysis – Internal Use of Results**
  - Compare alternative design options for company's own product or packaging
  - Compare new concept design with alternatives already in the marketplace to make a business development decision
- **Comparative Analysis – External Use of Results**
  - Provide science-based defense to public concerns or criticisms of a product's environmental performance compared to alternatives, including proposed legislation or bans
  - Use LCA results as the basis for marketing statements comparing a company's product with competing products

The choices that are made influence the rest of the LCA procedure. The definition of the goal and intended use will guide the practitioner in setting the scope and boundaries for the analysis, including the need for critical review and the type of critical review required (it may be important that an external expert takes on this task).

2.2 Components of a Well-Defined Study

Defining the goal and scope entails close communication between the LCA practitioner and the commissioner of the LCA study, at a minimum. The commissioner's role is typically to state the intent of conducting the LCA by explaining why they want the assessment to be conducted. The practitioner's job is then to define and present the appropriate methodological choices. Together, they develop the goal and scope that

Example Life Cycle Assessment Goals

Private Sector

The private sector is incorporating LCA in many applications including various aspects of products throughout design and development; manufacturing; marketing; use and reuse; and disposal and end of life management. Reasons for commissioning an LCA include the following:

- Establishing a baseline of overall environmental impact to identify environmental "hotspots."
- Identifying possible opportunities for improvement across the product life cycle.
- Comparing alternative manufacturing processes or supply chains to identify potential tradeoffs.
- Determining the environmental preferability between alternative product choices.
- Improving products through continuous improvement set often with concrete reduction targets (e.g., successor product must be X% less impactful than its predecessor while providing comparable performance).

Public Sector

In general, governments and the public sector lag behind the private sector in terms of embracing LCA as a tool for supporting decision making. However, there are many measures that can be implemented to use LCA results in public policy making. This can occur at multiple levels and lead to an environment that allows life cycle thinking help set the course towards a greener, more environmentally sustainable economy, including:

- Informing government programs and prioritizing their activities.
- Establishing consistent policies across consumers, producers, suppliers, retailers and waste managers.
- Establishing consistent policies and policy goals such as harmonizing regulations, voluntary agreements, taxes and subsidies.
- Introducing policies that appropriately support take-back systems to strengthen resource conserving-based economies.

is most likely to deliver results in line with the "why" question and provide data and information that are helpful to the commissioner of the study.

It is not unusual for the LCA commissioner to be rather vague initially and state their goal in very general terms, such as "We want to do an LCA on our products/production" or "We want to know the environmental strengths and weaknesses of this product" (Baumann and Tillman 2004). It is necessary to transform such general ideas into a specific purpose in order to adequately guide methodological choices which will result in useful results. This transformation may be done iteratively throughout the course of the study.

The end result should be an unambiguously stated goal with the reasons for carrying out the study. In defining the scope of the study, the following items should be considered and clearly described:

- the functions of the system(s) being studied;
- the functional unit;
- the system boundaries;
- allocation procedures;
- impact assessment methodology and interpretation approach to be used; and
- data needs.

Each of these components is explored separately in the following sections.

2.2.1 System Function

An essential aspect of goal and scope definition is the function and functional unit. Since any system is a collection of processes connected by flows of intermediate products, the system is defined by the function it provides. It can also be defined by the performance characteristics of a product.

For example, a system function to study a compact fluorescent light bulb (CFL) may be defined by the production, use and disposal of a single bulb or a set amount of the bulb, such as 1,000 bulbs. While the results of a study defined in this way would help identify environmental hotspots, a different function is needed if a comparison to a competing product is desired. To compare CFL to an incandescent light bulb, a more appropriate function would be performance based, i.e. the amount of light needed illuminate a room (such as 15 square meters with 1,000 lumen for 1 hour).

The function, in turn, determines the scope of the study which sets the boundaries and determines which unit processes will be included. And, because the system is a physical system it obeys the law of conservation of mass and energy. Mass and energy balances are the goal of a life cycle inventory and perform a useful check on the completeness and validity of the data.

2.2.2 Functional Unit

The functional unit provides a reference to which the inputs and outputs are related. That is, the selection of inputs and outputs to model the system are based on the functional unit. The reference flow is then used to calculate the inputs and outputs of the system (see box).

As already stated, most LCAs are performed to compare two or more systems. Comparisons between systems are made on the basis of the same function, quantified by the same functional unit in the form of their reference flows. The functional unit determination is important for both calculating flows between unit processes as well as for ensuring comparability between systems. Comparability is particularly critical when different systems (versus a baseline study or evaluating modifications to one system) are being assessed in order to ensure that comparisons are made on a common basis.

For example, in the *function* of drying hands in a public facility, a paper towel system is being compared to an air-dryer system. The selected *functional unit* may be expressed

Functional Unit versus Reference Flow

Functional unit is a quantified description of the service provided by the product system. It is determined by the study goal to answer the question (concern) at hand.

Example functional unit: Covering 20 m2 (215 ft2) area of new wall with 98% opacity and 5 year durability.

A reference flow is a quantified amount of manufactured product necessary for a product system to deliver the performance described by the functional unit.

Example reference flow: 3.8 liters (1 gallon) of primer
1liter (0.25 gallon) of paint

in terms of the identical number of pairs of hands dried for both systems. For each system, it is necessary to determine the reference flows through the product system based on the functional unit, e.g., the average mass of paper used or the average volume of hot air required to dry hands each time. In cases like this where the use is small, it is common to multiply the product use in order to get to numbers that are easier to work with (for example, 1,000 times to dry hands).

Returning to the light bulb example, it is meaningless to compare an incandescent light bulb with an LED light bulb since these products have different life spans and perform considerably differently (see Table 2.1). The functional unit expresses the function of the products, and, thereby, offers a way to equalize differences in performance. A functional unit for analyzing lighting systems could thus better be phrased in terms of the function, for instance "lighting a standard room of 15 square meters with 1000 lumen for 1 hour." As LCA mathematically employs a linear calculation rule, the results will scale by choosing a numerically different functional unit (say, "lighting a standard room of 20 square meters with 800 lumen for 3 hours"), but the alternatives considered will scale up or down consistently, so this will not affect the conclusions.

In a comparative analysis, the functional unit must take into account differences in the properties of the product, such as strength or durability, or differences in the use phase of the product. If the properties and performance of each of the systems analyzed is the same, the systems can be compared on a one-to-one basis. In cases where the reference flows are immediately equivalent, no adjustment is necessary. However, sometimes differences in the systems must be taken into consideration when defining

Table 2.1 Performance Characteristics of LED, Incandescent and CFL Light Bulbs.

	Light emitting diodes (LEDs)	Incandescent light bulbs	Compact fluorescents (CFLs)
Lumens	**Watts**	**Watts**	**Watts**
450	4-5	40	9-13
800	6-8	60	13-15
1,100	9-13	75	18-25
1,600	16-20	100	23-30
2,600	25-28	150	30-55
Life Span (average)	**50,000 hours**	**1,200 hours**	**8,000 hours**
Watts of electricity used (equivalent to 60 watt bulb) LEDs use less power (watts) per unit of light generated (lumens). LEDs help reduce greenhouse gas emissions from power plants and lower electric bills	6-8 watts	60 watts	13-15 watts
Kilo-watts of Electricity used (30 Incandescent Bulbs per year equivalent)	329 KWh/yr.	3285 KWh/yr.	767 KWh/yr.

the functional unit and scope of the analysis. An adjustment may be necessary in order to provide an apple-to-apple comparison (ISO 14049 2000). For example, in studying a cotton sock to a cotton sock embedded with nanosilver (Meyer *et al* 2010), the function of the nanosilver embedded in the sock to provide anti-bacterial properties must be taken into account. This can be accomplished by assessing a product, such as a foot spray or powder, along with the plain cotton sock.

For shipping containers, the function is to deliver an undamaged quantity of goods. Therefore, for containers that have equivalent capacity and provide equivalent product protection, an appropriate functional unit might be defined as 1,000 container shipments. If the containers being compared are a single-trip container and a reusable container, it is important to consider all the differences involved in making an equivalent number of shipments in each type of container.

For the *single-use* container, each time a shipment is made, a container must be manufactured and shipped to the filler. At the end user destination, the emptied container must be managed by recycling, landfill disposal, or combustion.

For the *reusable* container, the number of containers that must be manufactured for a given number of shipments depends on many factors including the percentage of containers that are lost or stolen during use, damage rates, fate of damaged containers (repaired, recycled, or disposed), and lifetime uses for containers that remain in circulation. Although fewer reusable containers must be manufactured for a given number of shipments, there are additional impacts associated with reuse that are not required for disposable containers. Reusable containers must be backhauled to the filler or to an inspection point. Backhauling may be done on a truck that is already returning from a retail store to a distribution center, or a special backhaul trip may be required. Reusable containers may require cleaning or reconditioning before they can be returned to use, which adds environmental burdens for materials and energy used in the cleaning process. Depending on where the cleaning takes place, additional transportation may also be required.

There are other differences between systems that may involve a combination of physical differences and consumer behavior. Consider two ice cream cartons holding an equivalent volume of product. Carton A is cylindrical with a lift-off lid, while Carton B is rectangular, with a paperboard flap closure. Carton B can be packed more compactly in a store freezer, so that less shelf space is required than Carton A. Because Carton A occupies more freezer space per unit volume of ice cream, Carton A is allocated a larger share of the daily energy requirements for operating the store freezer. However, consumers tend to prefer Carton A's removable lid design over Carton B's paperboard flap closure. If consumer preference for the Carton A design translates into faster sales compared to Carton B, the reduced time in the retail freezer for Carton A may offset its additional freezer shelf space requirement. When consumer behavior is involved, it is advisable to conduct a sensitivity analysis unless data are available to reliably characterize actual consumer behavior.

2.2.2.1 Functional Unit in Modeling the Agri-Food Industry

Starting from these general considerations, valid for all products, it is possible to review the particular issues regarding the choice of the functional unit of an agri-food product. Some reviews of LCA studies of food and agricultural products have shown first of all, in most cases, the choice of functional unit is based on the mass or volume (Hospida

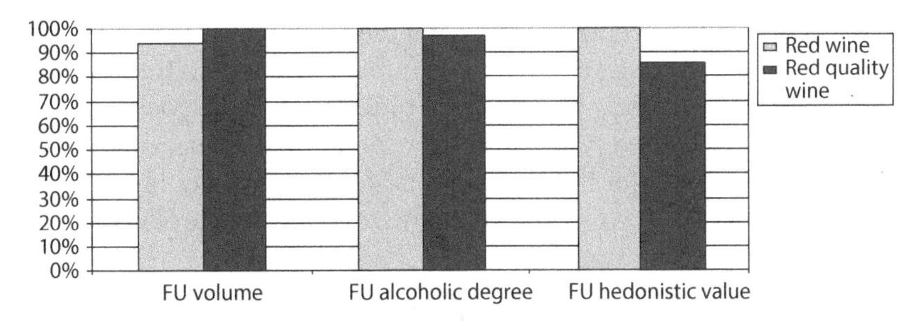

Figure 2.1 Abiotic Depletion Potential Characterisation of Two Wine Systems Assuming Volume, Alcoholic Degree and Hedonistic Value as Functional Unit (FU).

et al 2010; Roy *et al* 2009; Schau and Fet 2008). However, this is very often a limitation of the study, as food products derive from complex production activities which employ different technologies that make them not exactly comparable, even when belonging to the same type. In fact one of the main characteristics of food is to carry out various functions via its various nutrients. Therefore, it may be useful to choose a set of different functional units for the analyzed system in order to evaluate the variability of the results in terms of emissions (see Figure 2.1).

In food LCA methodology, food and packaging wastes are normally shown as an output from the system to either other systems (material recycling or energy recovery) or to Nature (deposits), where only the impacts of waste treatment (with some allocation factors between systems) are shown in each stage in the LCA model. Loss of product in each part of the chain can be compensated for in the reference flow if the functional unit has been defined as 1 kg of food being sold to or better utilized by the consumer. In such a case, product loss will be shown as increased volume of product being produced, packed and distributed, indicating higher environmental impacts from those earlier stages in the product chain, with a larger potential for improvement at those stages. However, the effectiveness loss is in fact at those stages where the food waste is generated, and should better be shown as an impact due to product loss at those stages, and not only as an impact of waste treatment. Defining the functional unit as 1 kg of product packed and distributed and effectively consumed by an average consumer makes it possible to compare different packaged systems with respect to efficiency along the distribution chain, and not consider only waste treatment impacts.

2.2.2.2 Functional Unit in Modeling Waste Management

The function of the system for an LCA of solid waste management alternatives is typically to manage waste. Therefore the functional unit can be defined as the management of a given quantity and composition of MSW disposed for a defined region under study. Alternatively, for a product LCA, the functional unit could be the "per ton" composition of waste disposed based on the discard of a product at the end of its useful life. If the goal of the LCA were to evaluate the overall environmental burdens associated with managing community "X's" annual waste generated, then the functional unit and formatting/presentation of results would likely be on a per annual tonnage of waste managed, and/or per ton of waste managed, basis.

If the goal of the LCA is to examine and compare alternatives for energy-from-waste to other sources of energy (e.g., electricity from fossil fuels), then the system function and functional unit would likely be different. For example, if the goal of the LCA were to evaluation the energy recovery potential for various waste management alternatives, then the functional unit and formatting/presentation of results would likely be on a per unit of energy produced basis (e.g., per 1000 Mwh of electricity produced)

2.2.3 Defining the System Boundaries (Scoping)

The essential feature of LCA in which it distinguishes itself from the analysis of an industrial or agricultural process is that it connects different unit process into a system. A flow diagram is a graphical representations of the system comprised of connected unit processes within a product system. This can be then represented in graphically in a flow diagram, such as the example in Figure 2.2.

As we can see, some unit processes are connected with one another in simple upstream-downstream connections, e.g., TV production is upstream connected to semi-conductor production. But there are also more complicated connections, e.g., electricity linking to different parts of the system, and recycling feeding back to production. Flow diagrams are in fact huge webs of interconnected unit processes. In the present era of digital databases, LCA studies can easily comprise several thousands of unit processes.

The scoping process must capture all stages and operations that are needed for the functional equivalence basis that has been selected for the analysis. Before making the decision to exclude a life cycle stage, the implications must be carefully considered. The scope definition further sets the main outline on a number of subjects that are discussed and further refined in more detail in the later phases. These include, among others:

- system boundaries;
- selection of impact categories;
- treatment of uncertainty.

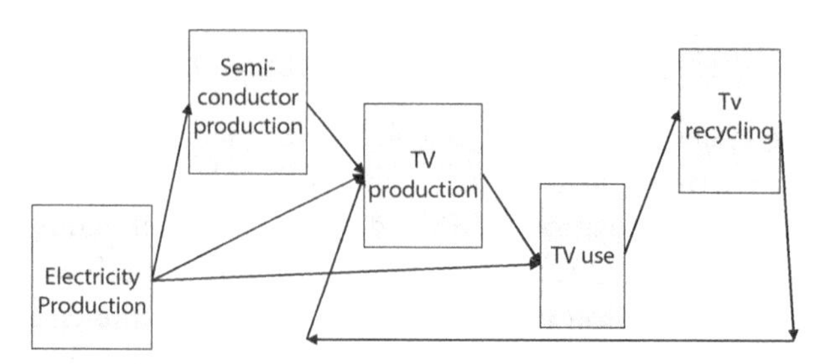

Figure 2.2 Fragment of a simplified flow diagram for an LCA on television (TV) sets. Because the purpose is to show how unit process are connected, only the flows from and to other unit processes are displayed; flows from and to the environment as well as transport, packaging, etc., are not displayed.

System boundaries must be defined in terms of the life cycle stages to be included in the analysis, the geographic and time boundaries of the analysis, and the flows and impact categories to be included.

The geographic boundaries of the system influence factors such as raw material sourcing, technology used, electricity grids, and transportation distances. Some materials have international raw material supply chains that are dependent on the geographic distribution of ores and other natural resources. Other processes may be specifically located to take advantage of material or energy supplies. For example, electricity-intensive alumina smelting operations are sited to take advantage of hydropower. Modeling the aluminum supply chain using average U.S. grid electricity for all the processes would result in a large overstatement of carbon dioxide emissions associated with smelting.

2.2.4 Co-Product Allocation

Many processes produce more than one useful output and are considered "multi-functional." This situation makes it necessary to use some method to divide or partition the process input and output flows among the useful outputs. The ISO standard outlines a hierarchy for addressing co-product allocation. The preferred approach involves dividing the unit process into its sub-processes. This is followed by system expansion to include the additional functions. Partitioning on the basis of physical parameters is then allowed, followed by the use of economic parameters. How co-product allocation is to be handled in an LCA should be clearly spelled out in the goal and scope phase. This topic is discussed in greater detail in Chapter 3 Life Cycle Inventory.

2.2.5 Impact Assessment

When scoping a project, it is important to define the impact categories that will be included in the results, as this choice influences the data collection requirements. A study scoped as a life cycle carbon footprint analysis will have very different data requirements than a study that includes a full set of life cycle impact indicators.

The impact assessment categories should link the potential impacts and effects to the entities that we aim to protect. The commonly-accepted areas of protection (AoP) are:

- Natural Resources
- Natural Environment
- Human Health
- and often also, Man-Made Environment

Multiple impact pathways originating from the LCI link emissions and extractions to impact category indicators. In practice, a category indicator is the outcome of a simplified model of a very complex reality, giving only an approximation of the quality status of the affected entity. Impact categories and corresponding indicators can be organized at two levels along the cause-effect chain: at a midpoint and at an endpoint level (Jolliet, Müller-Wenk *et al* 2004 and Bare and Gloria 2006). Figure 2.3 provides an example of a graphical representation of the midpoint-endpoint framework as proposed by the ILCD Handbook (JRC 2010).

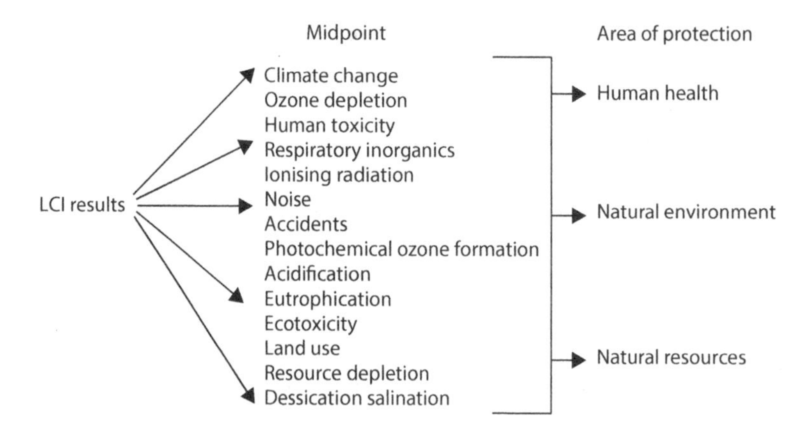

Figure 2.3 Relationship between midpoint impact categories and three areas of protection (adapted from JRC 2010).

2.3 Consequential LCA

LCA was initially developed to assess industrial systems related to consumer products. Later, there was a distinct shift in applying it to larger (regional and national) scales of industrial operations. LCA practitioners began making a distinction between how LCAs that accounted for stoichiometric-like relationships between physical flows to and from a product or process in an attributional style, to a ones that were more encompassing of the consequences of change in response to decisions, in a consequential LCA (Curran *et al* 2005). As a result, the process of system expansion (to avoid or deal with the allocation problem in multi-product systems) is an inherent part of consequential LCA studies. Ultimately, choosing between an attributional and a consequential LCA is decided by the defined goal of the study.

An example of the how the results of an attributional LCA and a consequential LCA can differ was presented by Searchinger *et al* (2008). In an attributional analysis of US corn-based ethanol, they calculated a 20% decrease in greenhouse gas emissions compared to conventional gasoline. However, in a consequential analysis to account for policy-driven increases in output, they predicted a 47% increase in emissions compared to gasoline, due to land use changes induced by higher prices of corn, soybeans and other grains from anticipated additional demand for corn starch for ethanol production.

Approaches for modeling consequential LCA are discussed in greater detail in Chapter 3 Life Cycle Inventory.

2.4 Carbon Footprint versus LCA

Assessing the carbon balance, also called carbon footprint, is of growing interest in environmental assessments of products. As is well known, the carbon cycle is the phenomenon by with which carbon circulates from the atmosphere to plants, animals, soil and back into the atmosphere. In food production, the crop is considered as an industrial process. This complicates the distinction between the natural system (biosphere) and the technical system (technosphere). The quantity of carbon sequestered during photosynthesis by the plant, for instance, has to be considered, but that same quantity of carbon will be emitted to the atmosphere during all the combustion processes, including the consumption of the food, human digestion, respiration, and metabolism activity. Consequently, if one accounts for the carbon sequestered, one has also to account for the carbon that will be emitted by the human body, as a consequence of the food consumption. Carbon dioxide emissions to atmosphere during human body respiration or, more in general carbon emission during digestion-excretion will counterbalance the biologic carbon sequestered by the plant from atmosphere. Human digestion and excretion remains the least studied life cycle stage of food products, it is comparable to the waste management stage for industrial products and its omission from LCA may compromise the results in identifying hot spots and opportunities for environmental improvement in the life cycle (Muñoz *et al* 2008; McLaren 2010).

For a long time, food LCA studies considered the carbon balance as net zero. Therefore, data about carbon sequestration and emissions along the life cycle of the product were not included in the boundaries of the study. Following this approach, however, some positive and negative factors of certain agricultural practices have not been properly considered. For example, organic farming or composting increases the soil organic matter, with positive consequences on biodiversity, but this has not been taken into account and no effect on the impact categories, such as global warming, was considered. The same considerations apply to the practices of intensive tillage operation that, on the contrary, deplete the soil of organic substances. Many studies, therefore, currently focus on the verification of the carbon balance, which is no longer zero, but can be modified depending on which effect overrides the other (sequestration or emission). Of course, the effect of sequestration prevails in the majority of studies that follow this approach; therefore, the total carbon balance is negative (good for the environment).

If the carbon balance refers to the effects described it is important not to report the carbon gain to only the annual cycle of the crop but it is necessary to consider the land use for 100 years, in line with the characterization factor of global warming. This is because many of the practices, that have a positive effect on carbon sequestration, have an incremental effect in carbon sequestration which is not infinite or continuous in time. For example, the transition from conventional farming to organic farming in the early years increases the organic matter content in the soil; however, this value settles for all subsequent years and does not increase further. The LCA should therefore consider the amount of carbon sequestered through agricultural practice and divide it by 100 years, so that the environmental gain is not assigned just to one year of observation. The calculation of the increase in the amount of dry biomass stably incorporated into the plant should follow the same approach.

2.5 Creating a Goal Statement

The objectives of an LCA are to examine system-wide effects on a cradle-to-grave basis, in order to assess all potential impacts to all media. Only through the consideration of the entire suite of issues can potential trade-offs be identified when systems are changed or a selection is made between choices. Out the outset, the goal and scope of an LCA study shall be clearly defined and consistent with the intended application.

LCA has several possible uses, including establishing a baseline of environmental impacts and forming the basis of eco-labeling, but identifying opportunities for improvement is a key application. Used with other information, the results of an LCA can be used to support decision making and provide the basis for achieving sustainability. This is the "begin with the end in mind" philosophy as promoted by Steven Covey. What do you want to get out of the LCA?

Below are example outcomes taken from a US Environmental Protection Agency (US EPA) guidance document on LCA (2006), which can help define the goals and scope of an LCA project:

- *Support broad environmental assessments* - The results of an LCA are valuable in understanding the relative environmental burdens resulting from evolutionary changes in given processes, products, or packaging over time; in understanding the relative environmental burdens between alternative processes or materials used to make, distribute, or use the same product; and in comparing the environmental aspects of alternative products that serve the same use.
- *Establish baseline information for a process* - A key application of an LCA is to establish a baseline of information on an entire system given current or predicted practices in the manufacture, use, and disposal of the product or category of products. In some cases, it may suffice to establish a baseline for certain processes associated with a product or package. This baseline would consist of the energy and resource requirements and the environmental loadings from the product or process systems that are analyzed. The baseline information is valuable for initiating improvement analysis by applying specific changes to the baseline system.
- *Rank the relative contribution of individual steps or processes* - The LCA results provide detailed data regarding the individual contributions of each step in the system studied to the total system. The data can provide direction to efforts for change by showing which steps require the most energy or other resources, or which steps contribute the most pollutants. This application is especially relevant for internal industry studies to support decisions on pollution prevention, resource conservation, and waste minimization opportunities.
- *Identify data gaps* - The performance of an LCA for a particular system reveals areas in which data for particular processes are lacking or are of uncertain or questionable quality. Inventory followed by impact assessment aids in identifying areas where data augmentation is appropriate for both stages.
- *Support public policy* - For the public policymaker, LCA can help broaden the range of environmental issues considered in developing regulations or setting policies.

- *Support product certification* - Product certifications have tended to focus on relatively few criteria. LCA, only when applied using appropriate impact assessment, can provide information on the individual, simultaneous effects of many product attributes.
- *Provide information and direction to decision-makers* - LCA can be used to inform industry, government, and consumers on the tradeoffs of alternative processes, products, and materials. The data can give industry direction in decisions regarding production materials and processes and create a better informed public regarding environmental issues and consumer choices.
- *Guide product and process development* - LCA can help guide manufacturers in the development of new products, processes.

As in any project, it's important to identify which of these – or other – goals and outcomes will define success. Following then, when reporting the results of an LCA, a clearly worded goal statement should help the reader quickly and precisely understand the question addressed and the main principles chosen in conducting the study. The goal statement should identify the following topics:

- the intended application;
- the reasons for carrying out the study;
- the intended audience to whom the results are to be communicated;
- whether the results are to be used in comparative assertions disclosed to the public.

The goal may be relatively simple and straightforward, for example, to assess the energy and greenhouse gas impacts associated with production of a single product, for internal use by the producer as a benchmark for evaluating future process improvements or design changes. In other cases the goal can lead to a very complex analysis, for example, a study with the goal of making public comparative claims about environmental performance of several competing products with variations in functional properties.

Also important is the need to specify geographic (spatial) and time (temporal) boundaries in order to guide data collection to be representative of the purpose of the location and time period of interest. Spatial boundaries, especially, are important because of differing environmental requirements (such as discharge limits) and consumer behaviors in different cities, states and countries. Also, physical realities differ by location, such as fresh water availability. Last but not least, generic or background data should be checked for spatial appropriateness, as supply chains and technologies can vary in different regions.

Time boundaries are important for similar reasons. Since life cycle inventories consist of large amounts of diverse data (i.e. they are sourced from different databases) it is unlikely that all data will have been collected within the relevant time period. In this case, the practitioner must evaluate if the data are still representative and usable or if new, more recent data are needed. Since most databases have a lag time before updating datasets, reference periods are usually set in the near past, rather than in the year in which the assessment is being conducted.

2.6 Preparing a Goal and Scope Document

At times it is beneficial to create an entire goal and scope document that clearly describes the essential aspects covered in the goal and scope definition phase, along with an introduction providing background details of the study. This is especially important if critical peer review is being conducted as the study progresses, and the goal and scope are part of the review process. There is no standard format for a goal and scope document to follow. The aim is for the goal and scope document to provide readers with a concise view of how the study is intended to proceed.

A hypothetical goal and scope document for a comparative LCA of Drinking Water Systems is presented in the appendix to this chapter. Based on information found in the

Example Goal Statement 1

The use of fossil fuels in automobiles is a major concern due the use of a non-renewable resource and contribution to greenhouse gas emissions. Thus, the use of bio-based feedstocks to make fuels, such as bioethanol, is a promising option to reduce the environmental impact of automotive transportation.

The goal of this consequential Life Cycle Assessment (LCA) is to identify whether it is beneficial from an environmental point of view to use bioethanol instead of traditional fuels in automobiles operated in the United States. The function was set at operating a typical passenger vehicle, averaging 24 miles per gallon, 12,000 miles in one year (2010). The results will be used to inform the public of the potential environmental consequences of bioethanol as an alternative fuel.

Example Goal Statement 2

The use of fossil fuels in automobiles is a major concern due the use of a non-renewable resource and contribution to greenhouse gas emissions. Thus, the use of bio-based feedstocks to make fuels, such as bioethanol, is a promising option to reduce the environmental impact of automotive transportation.

The goal of this attributional Life Cycle Assessment (LCA) is to identify the environmental impacts associated with the production, distribution and use of bioethanol. The function was set at producing 1,000 gallons of bioethanol from corn for use in a typical passenger vehicle operated in the United States in 2010. The results will be considered by fuel ethanol producers to identify potential opportunities for improving the fuel system.

full report, dated 22 October 2009, it was prepared in order to suggest how a goal and scope definition document would have looked if one had been prepared at the outset of the study. Goal and Scope Definition documents are rarely made publicly available, even after completion of a study. Their purpose is mainly for review, either internally or by an external review panel. The goal and scope are then documented in the final report. The drinking water systems report can be found on the Oregon DEQ website: http://www.deq.state.or.us/lq/sw/wasteprevention/drinkingwater.htm.

References

Bare JC and Gloria TP (2006) Critical Analysis of the Mathematical Relationships and Comprehensiveness of Life cycle Impact Assessment Approaches *Environ Sci Technol* 40(4): 1104-1113.

Baumann H and Tillman M (2004) The Hitch Hiker's Guide to LCA. Studentlitteratur, Lund, Sweden. ISBN 91-44-02364-2. 543 pp.

Curran MA, Mann M, Norris G (2005) The International Workshop on Electricity Data for Life Cycle Inventories *J Clean Prod* 13:853-862.

Jolliet O, Müller-Wenk R *et al* (2004) The LCIA Midpoint-Damage Framework of the UNEP/SETAC Life Cycle Initiative *Int J Life Cycle Assess* 9(6):394-404.

JRC (2010) Framework and Requirements for Life Cycle Impact Assessment (LCIA) Models and Indicators in ILCD Handbook - International Reference Life Cycle Data System, European Commission - Joint Research Center.

Hospido A, Davis J, Berlin J and Sonesson U (2010) A review of methodological issues affecting LCA of novel food products *Int J Life Cycle Assess* 15(1):44-52.

ISO (2000) Technical Report 14049:2000 Environmental management - Life cycle assessment - Examples of Application of ISO 14041 to Goal and Scope Definition and Inventory Analysis, International Standards Organization, Geneva, Switzerland.

Meyer D, Curran MA, and Gonzalez M (2010) An Examination of Silver Nanoparticles in Socks Using Screening-Level Life Cycle Assessment *J Nanopat Res* DOI 10.1007/s11051-010-0013-4.

Roy P, Nei D, Orikasa T, Xu Q, Okadome H, Nakamura N and Shiina T (2009) A Review of Life Cycle Assessment (LCA) on Some Food Products *Journal of Food Engineering* 90(1):1-10.

Schau E and Fet A (2008) LCA Studies of Food Products as Background for Environmental Product Declarations *Int J Life Cycle* Assess 13(3):255-264.

Searchinger T, Heimlich R, *et al* (2008) Use of U.S. Croplands for Biofuels Increases Greenhouse Gases through Emissions from Land-Use Change *Science* 319(5867):1238-1240.

US EPA (2006) Life Cycle Assessment: Principles and Practice, EPA/600/R-06/066, Office of Research and Development, Washington, DC, USA.

Appendix: Hypothetical Example of a Comparative, Attributional Life Cycle Assessment to Support Government Decision Making

Life Cycle Assessment of Drinking Water Systems: Bottled Water, Tap Water, and Home/Office Delivery (HOD) Water

Introduction

In 2009, the Oregon Department of Environmental Quality (DEQ) commissioned a life cycle assessment (LCA) study to compare alternative types of drinking water systems – water packaged in disposable bottles, tap water consumed from reusable drinking containers, and home/office delivery (HOD) water consumed from reusable drinking containers. The study was conducted for DEQ by ERG as an independent contractor.

The project at ERG was managed by Beverly J. Sauer, who served as primary life cycle analyst. Greg Schivley and Ann Marie Molen assisted with research tasks and development of the report appendices. Chris Dettore, a graduate student at the University of Michigan, provided assistance with research and contribution analysis tasks, with oversight by Dr. Greg Keoleian of the University of Michigan's Center for Sustainable Systems. The project was peer reviewed by an expert panel consisting of Beth Quay, an independent consultant with expert knowledge of bottling systems (serving as review chair), Dr. David Allen of the University of Texas, and David Cornell, an independent consultant with expert knowledge of PET container systems.

This annex summarizes the goal and scope of the study as described in the full report, dated 22 October 2009. It is intended to suggest how a goal and scope definition document would have looked if one had been prepared at the outset of the study. The report is publicly available and can be found on DEQ's website: http://www.deq.state.or.us/lq/sw/wasteprevention/drinkingwater.htm.

The findings and conclusions presented in the report are strictly those of ERG. ERG makes no statements nor supports any conclusions other than those presented in the report. ERG and DEQ are not responsible for writing or preparing the following recreation of a goal and scope document.

Background

Bottled water offers consumers a clean, potable supply of drinking water for consumption at home or away from home. Some disposable water bottles are recyclable, and lightweighting of bottles and bottled water packaging have reduced the amount of packaging waste associated with bottled water consumption. However, bottled water is frequently consumed at away from home locations where access to container recycling may be limited. In addition, while recycling of postconsumer bottles and packaging reduces consumption of virgin material resources, other resources are used and wastes created when packaging is manufactured and bottled water is transported.

Consumers have other drinking water options that do not involve disposable containers. These include consumption of tap water from a container that can be washed and reused many times, or consumption of water from a home/office delivery system with the water dispensed into a reusable drinking container. However, while reusable systems require less use and disposal of material, these systems require washing of containers between uses, and in the case of HOD systems, transportation of the containers to and from the filler. These processes incur environmental burdens that may be higher or lower than the burdens for disposable container systems.

Life Cycle Assessment (LCA) has been recognized as a scientific method for making comprehensive, quantified evaluations of the environmental benefits and tradeoffs for the entire life cycle of a product system, beginning with raw material extraction and continuing through disposition at the end of its useful life. This LCA evaluates the environmental burdens for disposable and reusable systems for delivering drinking water.

Purpose of the Study

This LCA was commissioned by the Oregon Department of Environmental Quality (DEQ) to evaluate the environmental implications of various systems for delivery and consumption of drinking water, including bottled water, tap water consumed from reusable containers, and home/office delivery (HOD) water consumed from reusable containers. The analysis includes water processing, production of containers and packaging materials, filling, transport, and end-of-life management of containers and packaging. The analysis also looks at transportation of bottled water imported from several foreign locations. The results are not intended to be used to represent specific brands of bottled water or reusable containers available in the marketplace.

Intended Use

The primary intended use of the study results is to inform DEQ about the environmental burdens and tradeoffs associated with various options for providing drinking water to consumers and behavioral choices of consumers. DEQ is also interested in better understanding the environmental burdens and tradeoffs of end-of-life management options (recycling, composting, landfilling, etc.). This analysis contains comparative statements about the results for the drinking water subscenarios analyzed. Because DEQ will make the results of this study, including comparative statements, publicly available, this report is being peer reviewed in accordance with ISO standards for life cycle assessment.

Systems Studied

The following types of drinking water systems are analyzed in this study.

Bottled water packaged in and consumed from individual disposable bottles:

- Virgin polyethylene terephthalate (PET) bottles (16.9 ounce, 8 ounce, and one liter)
- PET bottles with a mix of virgin and recycled content (16.9 ounce)
- Bottles made of virgin polylactide (PLA) resin derived from corn (16.9 ounce)
- Glass bottles with a mix of virgin and recycled content (12 ounce)

Tap water consumed from reusable containers:

- Virgin aluminum bottle with plastic closure (20 ounce)
- Virgin steel bottle with plastic closure (27 ounce)
- Virgin plastic bottle with plastic closure (32 ounce)
- Drinking glass with a mix of virgin and recycled content (16 ounce)
- Home/office delivery (HOD) water consumed from reusable containers
- Virgin polycarbonate bottles
- Virgin PET bottles
- Same reusable containers listed under the Tap system.

Within these three general drinking water scenarios, a number of subscenarios will be evaluated for variations in container sizes, weights, transportation distances, recycled content and recycling rates, and many other variables. Forty-eight subscenarios are identified: 25 bottled water subscenarios (20 for PET bottles, 4 for PLA, 1 for glass), 12 subscenarios for tap water consumption using a variety of reusable drinking containers, and 11 subscenarios for HOD water consumed from reusable containers. Of the bottled water subscenarios, five include long-distance transport of water from another country or the Eastern U.S. to Oregon.

Functional Unit

In a life cycle study, systems are evaluated on the basis of providing a defined function (called the functional unit). The function of each system analyzed in this report is to deliver drinking water to consumers. The functional unit selected for this analysis is delivering 1,000 gallons of drinking water to a consumer, including use of a bottle or reusable drinking container, and end-of-life management of the containers and packaging. To provide some perspective, 1,000 gallons is the amount of water a person would consume in about 5.5 years if they drank eight 8-ounce servings of water a day.

The functional equivalence is based on delivering drinking water that meets water quality standards set by the Food and Drug Administration (FDA), EPA, and state governments. The scope of the analysis does not include evaluating other differences in the quality of the water (e.g., taste, fluoride or mineral content, etc.) or temperature of the water, or any potential health impacts that may be associated with the use of specific water container materials. Carbonated and flavored waters are excluded.

The functional unit is 1,000 gallons of delivered water for several reasons:

- This basis produces results of a sufficient magnitude to be shown as whole numbers in the results tables and figures. Using a smaller unit, such as a liter of water, would produce results of a very small magnitude that would need to be shown in scientific notation.
- It is easier to understand reuse rates for 5-gallon HOD bottles when the functional unit is a multiple of the container volume (e.g., 1,000 gallons = 200 HOD bottle trips).
- Bottled water is typically packaged and purchased in multi-container cases, so again it makes sense to use a basis that is a multiple of the functional unit (1,000 gallons = 315 cases of 24 16.9 oz bottles) rather than a fraction of a purchasing unit (1 liter = two 16.9 oz bottles, equivalent to 1/12 of a case, or 0.083 cases).

Results shown on the basis of 1,000 gallons can easily be converted to any desired volume basis. For example, to convert results per 1,000 gallons to result per liter, first divide the 1,000 gallon results by 1,000 (to arrive at results on a per gallon basis), then divide the per gallon results by 3.8 liters per gallon to arrive at per liter results.

Scope and Boundaries

This study is a complete life cycle assessment (LCA) as defined in the ISO standards 14040 and 14044. As such, the study includes definition of goal and scope, life cycle inventory (LCI), life cycle impact assessment (LCIA), and interpretation of results.

The analysis includes all steps in the production of each drinking water container system, from extraction of raw materials through production of the materials used in the containers, fabrication of finished containers and closures, and transport to filling locations:

- Raw material extraction (e.g., extraction of petroleum and natural gas as feedstocks for plastic resins; growing corn used as a feedstock for polylactide resin, commonly referred to as PLA)
- Processing and fabrication steps to transform raw materials into containers and closures (water bottles, HOD bottles, reusable containers)
- Manufacture of materials used to package containers for retail shipment (corrugated trays, plastic film)
- Water treatment processes

- Container filling and washing operations (including industrial washing of HOD bottles and home washing of reusable drinking vessels)
- Distribution of filled containers
- Optional processes for chilling water
- End-of-life management of containers and packaging.

Treatment of municipal drinking water and additional processing steps used to purify bottled municipal water and natural water such as spring water are included in the analysis. Bottle filling and washing operations are included, as is production of secondary packaging used for shipment of filled containers, distribution of filled containers, washing of reusable containers, and end-of-life management of containers and associated packaging components. Various options for chilling water are also included in the model, including home refrigeration, use of ice, and HOD chiller units.

All washing of reusable personal drinking containers in this study is modeled based on use of a residential dishwasher, which is expected to be the most common method used by consumers for washing of these containers. Containers may also be hand-washed; however, water and detergent use for hand washing can vary widely based on the practices of individual consumers.

The scope of the study does not include analysis of scenarios for HOD and tap water consumed from disposable cups, nor does the study include scenarios in which disposable drinking water bottles sold filled with water were refilled by consumers and used as a reusable drinking container. Additional at-home purification of tap water, such as use of tap water filters, is not included in the scope of the analysis. The scope of the analysis does not include greenhouse gas effects of direct and indirect land use changes that may be associated with corn growing for PLA production.

Material Requirements

Once the LCI study boundaries have been defined and the individual processes identified, a material balance is performed for each individual process. This analysis identifies and quantifies the input raw materials required per standard unit of output, such as 1,000 pounds, for each individual process included in the LCI. The purpose of the material balance is to determine the appropriate weight factors used in calculating the total energy requirements and environmental emissions associated with each process studied. Energy requirements and environmental emissions are determined for each process and expressed in terms of the standard unit of output.

Once the detailed material balance has been established for a standard unit of output for each process included in the LCI, a comprehensive material balance for the entire life cycle of each product system is constructed. This analysis determines the quantity of materials required from each process to produce and dispose of the required quantity of each system component and is typically illustrated as a flow chart. Data must be gathered for each process shown in the flow diagram, and the weight relationships of inputs and outputs for the various processes must be developed.

Energy Requirements

The average energy requirements for each process identified in the LCI are first quantified in terms of fuel or electricity units, such as cubic feet of natural gas, gallons of diesel fuel, or kilowatt-hours (kWh) of electricity. The fuel used to transport raw materials to each process is included as a part of the LCI energy requirements. Transportation energy requirements for each step in the life cycle are developed in the conventional units of ton-miles by each transport mode (e.g. truck, rail, barge, etc.). Government statistical data for the average efficiency of each transportation mode are used to convert from ton-miles to fuel consumption.

Once the fuel consumption for each industrial process and transportation step is quantified, the fuel units are converted from their original units to an equivalent Btu value based on standard conversion factors.

The conversion factors have been developed to account for the energy required to extract, transport, and process the fuels and to account for the energy content of the fuels. The energy to extract, transport, and process fuels into a usable form is labeled precombustion energy. For electricity, precombustion energy calculations include adjustments for the average efficiency of conversion of fuel to electricity and for transmission losses in power lines based on national averages.

The LCI methodology assigns a fuel-energy equivalent to raw materials that are derived from fossil fuels. Therefore, the total energy requirement for coal, natural gas, or petroleum based materials includes the fuel-energy of the raw material (called energy of material resource or inherent energy). In this study, this applies to the crude oil and natural gas used to produce the plastic resins. No fuel-energy equivalent is assigned to combustible materials, such as wood, that are not major fuel sources in North America.

The Btu values for fuels and electricity consumed in each industrial process are summed and categorized into an energy profile according to the six basic energy sources listed below:

- Natural gas
- Petroleum
- Coal
- Nuclear
- Hydropower
- Other

The "other" category includes sources such as solar, biomass and geothermal energy. Also included in the LCI energy profile are the Btu values for all transportation steps and all fossil fuel-derived raw materials.

Environmental Emissions

Environmental emissions are categorized as atmospheric emissions, waterborne emissions, and solid wastes and represent discharges into the environment after the effluents pass through existing emission control devices. Similar to energy, environmental emissions associated with processing fuels into usable forms are also included in the inventory. When it is not possible to obtain actual industry emissions data, published emissions standards are used as the basis for determining environmental emissions.

The different categories of atmospheric and waterborne emissions are not totaled in this LCI because it is widely recognized that various substances emitted to the air and water differ greatly in their effect on the environment.

Atmospheric Emissions. These emissions include substances classified by regulatory agencies as pollutants, as well as selected non-regulated emissions such as carbon dioxide. For each process, atmospheric emissions associated with the combustion of fuel for process or transportation energy, as well as any emissions released from the process itself, are included in this LCI. The amounts reported represent actual discharges into the atmosphere after the effluents pass through existing emission control devices. Some of the more commonly reported atmospheric emissions are: carbon dioxide, carbon monoxide, non-methane hydrocarbons, nitrogen oxides, particulates, and sulfur oxides.

Waterborne Emissions. As with atmospheric emissions, waterborne emissions include all substances classified as pollutants. The values reported are the average quantity of pollutants still present in the wastewater stream after wastewater treatment and represent discharges into receiving waters. This includes both process-related and fuel-related waterborne emissions. Some of the most commonly reported waterborne emissions are: acid, ammonia, biochemical oxygen demand (BOD), chemical oxygen demand (COD), chromium, dissolved solids, iron, and suspended solids.

Solid Wastes. This category includes solid wastes generated from all sources that are landfilled or disposed of in some other way, such as incineration with or without energy recovery. These include industrial process- and fuel-related wastes, as well as the packaging components that are disposed when a container of product is emptied. Examples of industrial process wastes are residuals from chemical processes and manufacturing scrap that is not recycled or sold. Examples of fuel-related solid wastes are ash generated by burning coal to produce electricity, or particulates from fuel combustion that are collected in air pollution control devices.

LCI Practitioner Methodology Variation

This study follows the fundamental methodology for performing LCIs described in the ISO 14040 and 14044 standards. However, for some specific aspects of life cycle inventory, there can be variations in the methodology used by experienced practitioners. These areas include the method used to allocate energy requirements and environmental releases among more than one useful product produced by a process, the method used to account for the energy contained in material feedstocks, and the methodology used to allocate environmental burdens for postconsumer recycled content and end-of-life recovery of materials for recycling. LCI practitioners vary to some extent in their approaches to these issues. The following sections describe the approach to each issue used in this study.

Co-Product Credit

One unique feature of life cycle inventories is that the quantification of inputs and outputs are related to a specific amount of product from a process. However, it is sometimes difficult or impossible to identify which inputs and outputs are associated with individual products of interest resulting from a single process (or process sequence) that produces multiple useful products. The practice of allocating inputs and outputs among multiple products from a process is often referred to as "co-product credit"[1] or "partitioning."[2]

Co-product credit is done out of necessity when raw materials and emissions cannot be directly attributed to one of several product outputs from a system. It has long been recognized that the practice of giving co-product credit is less desirable than being able to identify which inputs lead to particular outputs. In this study, co-product allocations are necessary because of multiple useful outputs from some of the "upstream" chemical processes involved in producing the resins used to manufacture plastic packaging components.

Franklin Associates follows the guidelines for allocating co-product credit shown in the ISO 14044:2006 standard on life cycle assessment requirements and guidelines. In this standard, the preferred hierarchy for handling allocation is (1) avoid allocation where possible, (2) allocate flows based on direct physical relationships to product outputs, (3) use some other relationship between elementary flows and product output. No single allocation method is suitable for every scenario. How product allocation is made will vary from one system to another but the choice of parameter is not arbitrary. ISO 14044 section 4.3.4.2 states "The inventory is based on material balances between input and output. Allocation procedures should therefore approximate as much as possible such fundamental input/output relationships and characteristics."

Some processes lend themselves to physical allocation because they have physical parameters that provide a good representation of the environmental burdens of each co-product. Examples of various allocation methods are mass, stoichiometric, elemental, reaction enthalpy, and economic allocation. Simple mass and enthalpy allocation have been chosen as the common forms of allocation in this analysis. However, these allocation methods were not chosen as a default choice, but made on a case by case basis after due consideration of the chemistry and basis for production.

In the sequence of processes used to produce resins that are used in the plastic containers and closures, some processes produce material or energy co-products. When the co-product is heat or steam or a co-product sold for use as a fuel, the energy content of the exported heat, steam, or fuel is shown as an energy credit for that process. When the co-product is a material, the process inputs and emissions are allocated to the primary product and co-product material(s) on a mass basis. (Allocation based on economic value can also be used to partition process burdens among useful co-products;

---

[1] Hunt, Robert G., Sellers, Jere D., and Franklin, William E. **Resource and Environmental Profile Analysis: A Life Cycle Environmental Assessment for Products and Procedures.** Environmental Impact Assessment Review. 1992; 12:245-269.

[2] Boustead, Ian. **Eco-balance Methodology for Commodity Thermoplastics.** A report for The Centre for Plastics in the Environment (PWMI). Brussels, Belgium. December, 1992.

however, this approach is less preferred under ISO life cycle standards, as it depends on the economic market, which can change dramatically over time depending on many factors unrelated to the chemical and physical relationships between process inputs and outputs.)

In this study, corn grain is modeled as an input to production of PLA bottles. When corn grain is produced, corn stover (stalks and leaves) is coproduced. There are several ways in which corn stover can be managed. It may be left in the field to decompose, used for animal feed, or burned. In addition, there are some efforts to utilize corn stover as a source of biomass-derived energy. In this analysis, all of the corn growing burdens are allocated to the corn grain. The study used as the source of the corn growing data did not explicitly discuss the quantity of stover and whether it was treated as a co-product or as a waste; the implicit assumption is that the stover was neither allocated any co-product benefits nor assigned any waste management burdens, which would correspond with a scenario in which the stover is simply left in the field to decompose.

In the sequence of process steps used to convert corn into starch at a wet mill, coproducts corn gluten and corn oil are also produced. For each process step at the mill, the energy and emissions are allocated to corn starch and other coproducts on a weight basis.

Energy of Material Resource

For some raw materials, such as petroleum, natural gas, and coal, the amount consumed in all industrial applications as fuel far exceeds the amount consumed as raw materials (feedstock) for products. The primary use of these materials in the marketplace is for energy. The total amount of these materials can be viewed as an energy pool or reserve.

The use of a certain amount of these materials as feedstocks for products, rather than as fuels, removes that amount of material from the energy pool, thereby reducing the amount of energy available for consumption. This use of available energy as feedstock is called the energy of material resource (EMR) and is included in the inventory. The energy of material resource represents the amount the energy pool is reduced by the consumption of fuel materials as raw materials in products and is quantified in energy units.

EMR is the energy content of the fuel materials *input* as raw materials or feedstocks. EMR assigned to a material is *not* the energy value of the final product, but is the energy value of the raw material at the point of extraction from its natural environment. For fossil fuels, this definition is straightforward. For instance, petroleum is extracted in the form of crude oil. Therefore, the EMR for petroleum is the higher heating value of crude oil.

Once the feedstock is converted to a product, there is energy content that could be recovered, for instance through combustion in a waste-to-energy waste disposal facility. The energy that can be recovered in this manner is always somewhat less than the feedstock energy because the steps to convert from a gas or liquid to a solid material reduce the amount of energy left in the product itself.

The materials which are primarily used as fuels (but that can also be used as material inputs) can change over time and with location. In the industrially developed countries included in this analysis, these materials are petroleum, natural gas, and coal. While some wood is burned for energy, the primary uses for wood are for products such as paper and lumber. Similarly, some oleochemical oils such as palm oils can be burned as fuel, often referred to as "bio-diesel." However, as in the case of wood, their primary consumption is as raw materials for products such as soaps, surfactants, cosmetics, etc.

At this time, the predominant use of biomass crops is for food or material use rather than as an energy resource. However, biomass is increasingly being used as feedstock for fuels, e.g., corn-derived ethanol and soy-derived biodiesel. At some point in the future, the energy of material resource methodology may be applied to biomass resources as well as fossil resources.

Postconsumer Recycling Methodology

Some drinking water containers are recycled at end of life. Some containers also have recycled content. When material is used in one system and subsequently recovered, reprocessed, and used in another application, there is a reduction in the total amount of virgin material that must be produced to fulfill the two systems' material needs. However, there are different methods by which the savings in virgin material production and disposal burdens can be assigned to the systems producing and using the recovered material. Material production, collection, reprocessing, and disposal burdens can be allocated over all the useful lives of the material, or boundaries can be drawn between each successive useful life of the material.

Because the choice of recycling allocation methodology can significantly influence the LCI results, several approaches are explored in this analysis, including sharing the burdens for a given quantity of resin equally between multiple uses of the resin (Method 1), assigning the resin production burdens to the system first using the virgin resin (Method 2), or transferring the resin production burdens from the system first using the virgin resin to the system that uses the recovered resin (Method 3). In all cases, the allocated burdens include the energy of material resource embodied in the plastic material.

Each recycling approach used in this analysis is described in more detail in the sections below. In these descriptions, the system from which the material is recovered is referred to as the "producer" system, and the system utilizing recovered material is referred to as the "user" system. It should be noted that all recycling allocations are based only on the burdens for the *resin material* and do not include any allocation of the burdens associated with fabricating the resin into a bottle or any other product. Thus, there are no inherent assumptions about the product in which resin is used before or after the resin's use in the bottle system.

Method 1: Open-loop Allocation. The recycling methodology designated method 1 in this analysis is an open-loop allocation approach. In this approach, all environmental burdens associated with a quantity of recycled material are shared equally between the systems producing and using the material, resulting in reduced burdens for both systems. The producer and user systems share the burdens for virgin material production, collection, reprocessing, and disposal, so that both systems share equally in the benefits of recycling.

For bottles that contain recycled material, the recycled resin content of the bottle comes into the bottle system with half of its virgin production burdens (as well as half of the burdens for collecting and reprocessing the material and disposing of the material at end of life). The other half is allocated to the original product system that used the material, which is outside the boundaries of this analysis. For example, if a bottle had recycled content "r", the recycled material in the bottle would carry half of the burdens required to produce, collect, reprocess, and dispose of that material, or $r/2 * (V+PC+D)$, where "V" is virgin material production burdens, "PC" is postconsumer collection and reprocessing burdens, and "D" is disposal burdens. The virgin percentage of the bottle would carry full burdens for material production and disposal, or $(1-r)*(V+D)$. Adding these together, the total virgin production burdens allocated to the recycled content bottle are $(r/2)*V + (1-r)*V$, or $(1-r/2)*V$. Similarly, the material disposal burdens allocated to the recycled content bottle are $(r/2)*D + (1-r)*D$, or $(1-r/2)*D$. The collection and reprocessing burdens for the recycled content allocated to the bottle are $r/2*PC$.

A similar allocation approach is used for virgin bottles that are recycled after use. If "R" percent of virgin bottles are recycled at end of life, with half the virgin burdens for the bottle material going to a subsequent use outside the boundaries of the bottle system, then the virgin burdens allocated to the bottle system for the recycled bottles are $R/2*(V+ PC+D)$ for the bottles that are recycled + $(1-R)*(V+D)$ for the material in the bottles that are not recycled. The total virgin production burdens allocated to the bottle are $(R/2)*V+ (1-R)*V$, or $(1-R/2)*V$, the allocated disposal burdens are $(R/2)*D + (1-R)*D$, or $(1-R/2)*D$, and the collection and reprocessing burdens are $R/2*PC$.

For bottles that contain recycled material *and* are recycled after use, allocation becomes more complicated. For an example of bottles with recycled content r *and* recycling rate R, the virgin burdens for the material *in* the bottle are $(1-r/2)*(V)$, as described above. Some of these burdens must then be allocated to the next use of the material, using the $(1-R/2)$ allocation. The net virgin burdens assigned to the bottle system, taking into account both the recycled content and the postconsumer recycling rate, are $(1-r/2)*V*(1-R/2)$. The allocated disposal burdens are $(1-r/2)*D*(1-R/2)$. The share of recycling burdens allocated to the bottle system is $r/2*PC*R/2$.

No further projections are made about the fate of the material after the end of its recycled use. For example, if a product made from recycled bottle material is subsequently recycled at the end of its life, then the material would have three uses rather than two. This analysis uses a conservative approach and takes into account only the *known* number of useful lives of the bottle material (i.e., one prior use for recycled material used in bottles that have recycled content; one subsequent use for bottle material that is recycled at end of life).

The other two recycling approaches are less complicated to model, as they draw boundaries between successive lives of the material, with burdens for specific steps allocated to either the producer system or the user system. When postconsumer material from one system is used in a second system, different perspectives can be taken as to whether the producer or user system deserves the credit for the reductions in virgin material production and material disposal due to recycling.

Method 2: User Credit Allocation. Recycling methodology 2 can be called the user credit method. In this approach the boundaries between successive uses of the material are drawn so that the system *using* the recycled material gets the credit for avoiding production of more virgin material. In method 2, all virgin material burdens for initially producing material are allocated to the first system using the material (e.g., a virgin water bottle), and the next system using the recovered material (resin from recovered bottles) takes all the burdens for collection and reprocessing of the material, as well as the burdens for disposing of the material (unless it is recycled again after use in the second system). The benefit to the producer system (in this example, the bottle system) is limited to avoided disposal burdens for the material that goes on to the secondary user. Using the same variables as above, the allocations are as follows:

For a bottle with recycled content r and recycling rate R, the virgin material production burdens assigned to the bottle are (1-r)*V, the recycling burdens are r*PC, and the disposal burdens are (1-R)*D.

Method 3: Producer Credit Allocation. Recycling method 3 can be referred to as the producer credit method. In this approach, the system *generating* the recovered material gets the credit for avoiding the need to produce more virgin material. Because the material is not disposed but goes on to a subsequent use, the producer system is assigned burdens for collecting and reprocessing the material in order to deliver it to the next user (in lieu of the burdens that would otherwise be incurred for disposing of the material). The virgin burdens for producing the material and the burdens for disposing of the material are transferred to the next system using the material, which may in turn pass these burdens on to a subsequent use if that product is recovered and recycled at end of life. Using the same variables as above, the allocations are as follows: For a bottle with recycled content r and recycling rate R, the virgin material production burdens assigned to the bottle are V*(1-R), the recycling burdens are R*PC, and the disposal burdens are D*(1-R).

System Expansion. Another approach that can be used to allocate burdens for coproducts or recycled products is system expansion, in which credit is given for a product or material that is displaced by the product or material of interest. In order to use system expansion, it is important to know the specific application that is being displaced, as different uses of material have different reprocessing requirements and different fabrication requirements. As noted previously, the recycling allocations in this analysis are applied only to the burdens associated with the resin material. The recycling allocations do not include additional processing to prepare the resin for a specific end use or fabricate it into a specific product (e.g., a food-grade application or production of carpet fiber) before or after its use in the bottle system, nor were any assumptions made about the previous or subsequent products in which the bottle resin would be used. The recycling burdens in this study are based on collection and mechanical recycling of PET bottles into "generic" clean flake, and not on displacement of any specific product.

Life Cycle Inventory Data

The accuracy of the study is directly related to the quality of input data. Data necessary for conducting this analysis are separated into two categories: process-related data and fuel-related data.

Process Data

Methodology for Collection/Verification. The process of gathering data is an iterative one. The data-gathering process for each system begins with a literature search to identify raw materials and processes necessary to produce the final product. The search is then extended to identify the raw materials and processes used to produce these raw materials. In this way, a flow diagram is systematically constructed to represent the production pathway of each system. Each process identified during the construction of the flow diagram is then researched to identify potential industry sources for data.

Confidentiality. Franklin Associates takes care to protect data that is considered confidential by individual data providers. This can be done by aggregating data with data sets from other sources for the same unit process or aggregating the data with other sequential life cycle unit processes. The appendices for this report (a separate document) present all data sets used in this analysis at the maximum level of detail possible while still protecting confidentiality.

Objectivity. Each unit process in the life cycle study is researched independently of all other processes. No calculations are performed to link processes together with the production of their raw materials until *after* data gathering and review are complete. This allows objective review of individual data sets before their contribution to the overall life cycle results has been determined. Also, because these data are reviewed individually, assumptions are reviewed based on their relevance to the process rather than their effect on the overall outcome of the study.

Fuel Data

When fuels are used for process or transportation energy, there are energy and emissions associated with the production and delivery of the fuels as well as the energy and emissions released when the fuels are burned. Before each fuel is usable, it must be mined, as in the case of coal or uranium, or extracted from the earth in some manner. Further processing is often necessary before the fuel is usable. For example, coal is crushed or pulverized and sometimes cleaned. Crude oil is refined to produce fuel oils, and "wet" natural gas is processed to produce natural gas liquids for fuel or feedstock.

To distinguish between environmental emissions from the combustion of fuels and emissions associated with the production of fuels, different terms are used to describe the different emissions. The combustion products of fuels are defined as combustion data. Energy consumption and emissions which result from the mining, refining, and transportation of fuels are defined as precombustion data. Precombustion data and combustion data together are referred to as fuel-related data.

Fuel-related data are developed for fuels that are burned directly in industrial furnaces, boilers, and transport vehicles. Fuel-related data are also developed for the production of electricity. These data are assembled into a database from which the energy requirements and environmental emissions for the production and combustion of process fuels are calculated.

Energy data are developed in the form of units of each primary fuel required per unit of each fuel type. For electricity production, federal government statistical records provided data for the amount of fuel required to produce electricity from each fuel source, and the total amount of electricity generated from petroleum, natural gas, coal, nuclear, hydropower, and other (solar, geothermal, etc.). In this study, the Oregon grid is used to model electricity used for processes taking place in Oregon. Literature sources and federal government statistical records provided data for the emissions resulting from the combustion of fuels in utility boilers, industrial boilers, stationary equipment such as pumps and compressors, and transportation equipment. Because electricity and other fuels are required in order to produce electricity and primary fuels, there is a complex and technically infinite set of interdependent steps involved in fuel modeling. An input-output modeling matrix is used for these calculations.

Trip Allocation for Purchases of Bottled Water

Unlike consumption of tap water, which requires no travel on the part of the consumer, and consumption of HOD water, which is delivered by a truck used specifically for this purpose, bottled water is most often picked up by the consumer on an outing that may have several purposes. The consumer is likely to run more than one errand on the same outing, and it is also likely that additional items will be purchased at the same location when the consumer purchases bottled water.

This analysis uses a modeling approach that is based on bottled water being purchased one case at a time, with 24 bottles per case. The number of trips required to purchase 1,000 gallons of water depends on the volume of water in an individual bottle and the number of bottles in the case, both of which can be varied in the model. Each time a trip is made to purchase water, it is assumed that the case of water is purchased on an outing that includes one other errand in addition to the stop where water is purchased. The round-trip distance from the consumer's home to the purchasing location is scaled up to account for the additional distance traveled to include the second stop (home to stop 1, stop 1 to stop 2, and stop 2 back to home). The overall distance traveled is divided by two to allocate half to each stop made.

Furthermore, it is reasonable to assume that any item purchased on a trip to a grocery or other retail store could warrant an individual trip to the store if the item were not purchased together with other items as part of a combined purchase. Therefore, the burdens for making the stop at the store can be allocated over the number of items purchased. For example, if 25 items are purchased on a trip to a store, each item would be allocated 4% of the burdens for making the stop at the store. For purchasing bottled water on a two-errand outing, most modeling scenarios in this analysis use a trip allocation of 4 percent, although one scenario models a two-errand trip in which only water is purchased on the stop at the grocery store, so that 100 percent of the burdens for that stop are allocated to water. The 25-item purchase is an estimate by the LCA practitioner, since no data were readily available for consumer purchasing patterns on an individual shopping trip basis.

In addition to allocating a portion of the total vehicle fuel use to bottled water, the analysis also accounts for the marginal increase in the loaded vehicle weight due to a case of water and the associated slight decrease in fuel economy over the distance the water is transported from store to home. The baseline fuel economy used for the consumer vehicle was 19.9 miles per gallon.[3]

System Components Not Included

The following components of each system are not included in this LCI study:

Water Use. There is currently a lack of water use data on a unit process level for life cycle inventories. In addition, water use data that are available from different sources do not use a consistent method of distinguishing between consumptive use and non-consumptive use of water or clearly identifying the water sources used (freshwater versus saltwater, groundwater versus surface water). A recent article in the International Journal of Life Cycle Assessment summarized the status and deficiencies of water use data for LCA, including the statement, "To date, data availability on freshwater use proves to be a limiting factor for establishing meaningful water footprints of products."[4] The article goes on to define the need for a standardized reporting format for water use, taking into account water type and quality as well as spatial and temporal level of detail.

Because of the lack of complete and consistent data on water use for raw material and intermediate unit processes, Franklin Associates' LCI database does not currently include water use. In this analysis, wastewater quantities are estimated only for water treatment processes and container washing operations.

Capital Equipment. The energy and wastes associated with the manufacture and installation of capital equipment and infrastructure are not included. This includes equipment to manufacture buildings, motor vehicles, and industrial machinery, and the installation of water distribution piping. The energy and emissions associated with such capital equipment generally, for 1,000 pounds of materials, become negligible when averaged over the millions of pounds of product manufactured over the useful lifetime of the capital equipment.

Space Conditioning. The fuels and power consumed to heat, cool, and light manufacturing establishments are omitted from the calculations in most cases. For manufacturing plants that carry out thermal processing or otherwise consume large amounts of energy, space conditioning energy is quite low compared to process energy. Energy consumed for space conditioning is usually less than one percent of the total energy consumption for the manufacturing process. This assumption has been checked in the past by Franklin Associates staff using confidential data from manufacturing plants. In this analysis, bottled water purchased in retail stores has not been assigned any share of the store's general space conditioning energy.

Support Personnel Requirements. The energy and wastes associated with research and development, sales, and administrative personnel or related activities have not been

---

[3] Average fuel economy for Oregon personal vehicles according to information provided by Oregon DEQ.

[4] Koehler A (2008) Water Use in LCA: Managing the Planet's Freshwater Resources *Int J Life Cycle Assess* 13:451-455.

included in this study. Similar to space conditioning, energy requirements and related emissions are assumed to be quite small for support personnel activities.

Miscellaneous Materials and Additives. Selected materials such as catalysts, pigments, or other additives which individually account for less than one percent by weight of the net process inputs are typically not included in the assessment unless inventory data for their production are readily available or there is reason to believe that these additives have environmental impacts that are very high in relation to their mass.

Rebound Effect. The analysis does not include any analysis of the environmental impacts of changes in consumer behavior that may be associated with choosing one water delivery system over another. For example, if consumers choose to drink tap water rather than purchasing bottled water, they may choose to save or invest the money that they do not spend on bottled water, or they may choose to spend the money on a different item or activity. Conversely, if consumers purchase bottled water, this will reduce the money they have available to spend on other items and activities. Alternative purchased items or activities may have environmental impacts that are greater or lesser than the impact of purchasing bottled water. It is beyond the scope of this analysis to make projections about the environmental impacts of alternative uses of consumers' spending dollars that are currently used to purchase bottled water.

Data Sources

Data from credible published sources or licensable databases will be used wherever possible in order to maximize transparency. Wherever possible the study uses Oregon-specific data and assumptions For processes and materials where reliable current published data were not available, data sets from Franklin Associates' United States industry average database will be used. This database has been developed over a period of years through research for many LCI projects encompassing a wide variety of products and materials. Another advantage of the database is that it is continually updated. For each ongoing LCI project, verification and updating is carried out for the portions of the database that are accessed by that project.

Data Quality Goals for This Study

ISO standard 14044:2006 states that "Data quality requirements shall be specified to enable the goal and scope of the LCA to be met." Data quality requirements include time-related coverage, geographical coverage, technology coverage, and more. The data quality goal for this study was to maximize transparency by using life cycle data from credible publicly available sources to the extent possible, and to model all systems to reflect Oregon-specific conditions and practices, where appropriate. Where publicly available life cycle data were not available, processes and materials in this study were modeled based on Franklin Associates' LCI database. The quality of individual data sets vary in terms of age, representativeness, measured values or estimates, etc.; however, all materials and process data sets used in this study will be thoroughly reviewed for accuracy and currency and updated to the best of our capabilities for this analysis.

Data Accuracy

An important issue to consider when using LCI study results is the reliability of the data. In a complex study with literally thousands of numeric entries, the accuracy of the data and how it affects conclusions is truly a complex subject, and one that does not lend itself to standard error analysis techniques. Techniques such as Monte Carlo analysis can be used to study uncertainty, but the greatest challenge is the lack of uncertainty data or probability distributions for key parameters, which are often only available as single point estimates. However, the reliability of the study can be assessed in other ways.

A key question is whether the LCI profiles are accurate and study conclusions are correct. The accuracy of an environmental profile depends on the accuracy of the numbers that are combined to arrive at that conclusion. Because of the many processes required to produce each container or packaging material, many numbers in the LCI are added together for a total numeric result. Each number by itself may contribute little to the total, so the accuracy of each number by itself has a small effect on the overall accuracy of the total. There is no widely accepted analytical method for assessing the accuracy of each number to any degree of confidence. For many chemical processes, the data sets are based on actual plant data reported by plant personnel. The data reported may represent operations for the previous year or may be representative of engineering and/or accounting methods. All data received are evaluated to determine whether or not they are representative of the typical industry practices for that operation or process being evaluated.

There are several other important points with regard to data accuracy. Each number generally contributes a small part to the total value, so a large error in one data point does not necessarily create a problem. For process steps that make a larger than average contribution to the total, special care is taken with the data quality.

There is another dimension to the reliability of the data. Certain numbers do not stand alone, but rather affect several numbers in the system. An example is the amount of material required for a process. This number will affect every step in the production sequence prior to the process. Errors such as this that propagate throughout the system are more significant in steps that are closest to the end of the production sequence. For example, changing the weight of an input to the final fabrication step for a plastic component changes the amounts of resin inputs to that process, and so on back to the quantities of crude oil and natural gas extracted.

In addition to the accuracy of the underlying data sets used to model each unit process, an added dimension to the drinking water analysis is the unlimited possibilities for variations in container weights, recycled content, fabrication energy, transportation distances, consumer use behavior, etc. for the drinking water systems studied. Because of this, the life cycle model was set up as a dynamic model capable of evaluating a wide range of user-defined scenarios. The program TopRank will also be used to evaluate the sensitivity of results to variations in individual modeling parameters.

Geographic Scope

Data for foreign processes are generally not available. This is usually only a consideration for the production of oil that is obtained from overseas. In cases such as this, the energy requirements and emissions are assumed to be the same as if the materials originated in the United States. Since foreign standards and regulations vary from those of the United States, it is acknowledged that this assumption may introduce some error. Transportation of crude oil used for petroleum fuels and plastic resins is modeled based on the current mix of domestic and imported crude oil used.

Other processes in this analysis modeled as occurring outside the United States include production of virgin aluminum and steel reusable drinking containers and the processing and bottling of water imported from several countries. Fabrication of the bottles used to package imported water was assumed to occur in the country in which the water was bottled. Recovered PET bottles are assumed to be exported to China for recycling, so PET resin production emissions are based on the U.S. grid, while credits for recycled resin are based on PET production using the Chinese electricity grid. (Recovered metals, glass, and corrugated were assumed to be recycled in the U.S.) For processes occurring outside the U.S., U.S. process energy requirements are used, but production of process electricity was modeled based on that country's electricity grid.

The following table summarizes the model settings used for the three example scenarios. For each drinking water system, the example scenario represents only one of the many combinations of parameters that can be modeled for each of the drinking water systems and is not meant to be interpreted as the most likely or most representative scenario for that system. Parameters that are modeled consistently for all systems (e.g., wastewater treatment) are not shown in the table.

Selection of Subscenarios

48 subscenarios meet the following goals:

- To capture scenarios that are believed to best represent typical practices
- To demonstrate "best case" or "worst case" scenarios for selected systems to see if results for the different drinking water systems (bottled, tap, HOD) overlap at practical extremes
- To explore compounding or offsetting effects of simultaneous variations in key parameters within systems
- To identify parameters that have a large effect on results
- To identify parameters that do not have a large effect on results at any level.

Because the results for the example systems showed higher results for the bottled water system, in most cases the selected subscenarios use conservative baseline estimates or assumptions for the bottled water system and less favorable baseline assumptions (e.g., one-year useful life, washing container after each use) for the reusable tap and HOD systems, to see if overlap is expected within the ranges of parameters that could occur for the different systems.

Life Cycle Impact Assessment

The U.S. EPA's TRACI methodology was selected as the impact assessment methodology to be used, since it was developed to represent U.S. conditions (e.g., for fate and transport of chemical releases).

Table 1 Modeling Parameters for Example Drinking Water Systems.

	Bottle Water	**Tap/Reusable**	**HOD/Reusable**
Bottle material	PET	Reusable virgin aluminium drinking container made in Switzerland	Reusable virgin aluminium drinking container made in Switzerland
Container capacity	16.9 oz	20 oz	20 oz
Container Weight	13.3 g	100 g	100 g
Recycled content	0%	0%	0%
Times filled per day		1	1
Years use before disposal or recycling		1	1
Days use of reusable drinking contasiner before washing in home dishwasher		1	1
Reusable container washing in home dishwasher with high or low water use		high water use	high water use
HOD container material			Polycarbonate
HOD container weight			750 g
Lifetime uses			40
Water in bottle	OR purified municipal water with reverse osmosis, ozone treatment, and UV	Oregon tap water (municipal water with no additional purification steps)	OR purified municipal water with reverse osmosis, ozone treatment, and UV
Transport of empty single-serve water bottle from off-site molder to filloer	0 miles (molded at filling location)		
Single-serve bottles rinsed before filling	no		

Distance from filler to store	50 mi		
Distance from store to home	5 mi		
Personal vehicle fuel usage allocated to purchased water	4% (e.g., 1 of 25 items purchased)		
Miles on HOD distribution route			75 miles
Chilling	none	none	HOD chiller unit
Container recycling (methodology 1)	62%	0%	0% recycling of aluminium bottle, 100% recycling of HOD container
Recycling of corrugated packaging	76%		

Chapter 2 Exercises

1. Defining the Goal

   An LCA starts with the definition of the intended goal and scope of the study. The functional unit provides a reference to which the inputs and outputs are related. Procter & Gamble has conducted LCAs on several of their product lines. The figure below shows the results of energy usage from a life cycle perspective.

   a) Write a goal statement for a study including a laundry detergent product. Consider the life cycle from different perspectives:
      - The product manufacturer
      - The consumer
      - The washing machine manufacturer

      In the goal statement, state whether:
      - The results are to be used internally for hotspot identification OR externally for making an environmental claim;
      - The goal is to provide a baseline of a single product system OR to compare two competing alternatives.

   b) How do these choices affect the study approach?

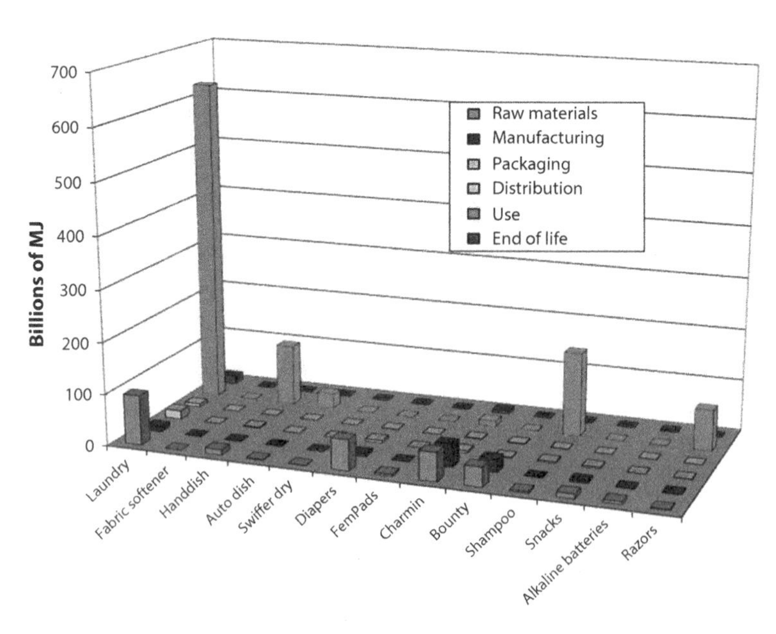

Figure E.1 Results of multiple LCAs show the different Cumulative Energy Demand (billions of MJ) of 13 product types sold in North America and Western Europe, and focuses attention on where innovation would be most helpful to reduce energy use.

2. Defining the Functional Unit
   For the product selected in Exercise 1, define the function of the system and the amount of product (the Functional Unit) needed to fulfill the function and the study's goal. This information should be included in the goal statement.
3. Defining the System Boundaries
   Draw the (cradle to grave) system flow diagram for laundry detergent.
   How would the results of the study differ if the boundaries had been set to evaluate only production (cradle to gate), assuming the use of the product was the same and excluded from study, rather than cradle to grave? What cut-off assumptions can be made?
4. LCA of Processes and Activities, too.
   LCA is described as being applicable not only to products but also to industrial processes and activities.
   a) Develop a goal statement for a system other than a consumer product, such as waste management or a water treatment process, site remediation activity, organic farming, lawn care, etc.
   b) Draw the system with defined boundaries.
5. Consequential LCA
   The figure below depicts flow diagrams for three motor oil systems (virgin oil, re-refined oil, bio-based oil). The original study was conducted as an attributional LCA. What additional considerations and data would be needed to conduct a consequential LCA of these systems?

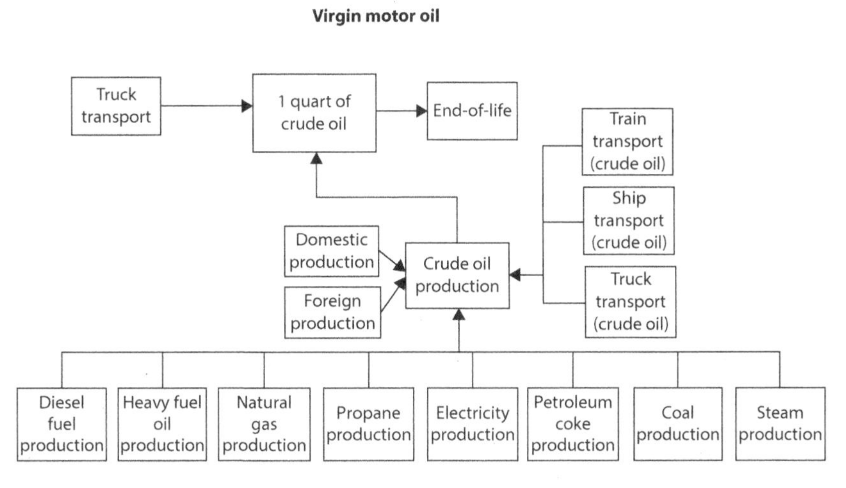

Figure 1 Flow diagram for production and use of virgin motor oil.

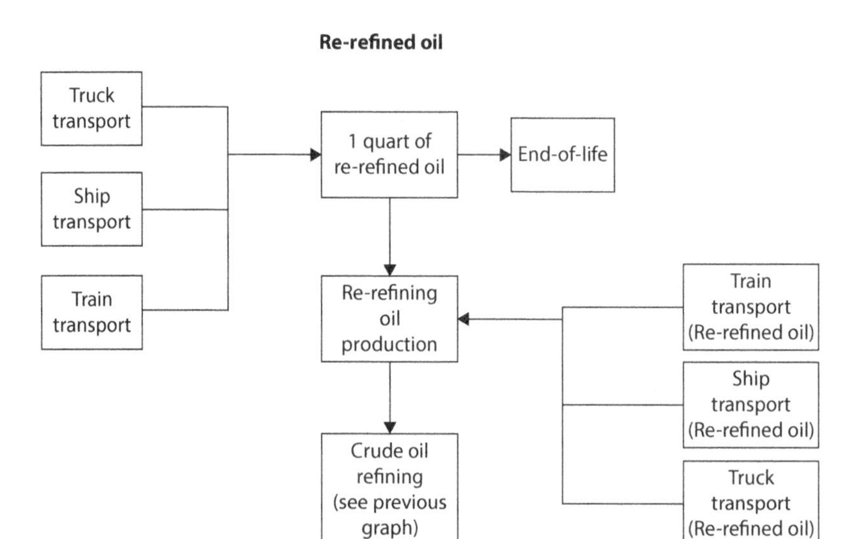

Figure 2 Flow diagram for production and use of re-refined motor oil.

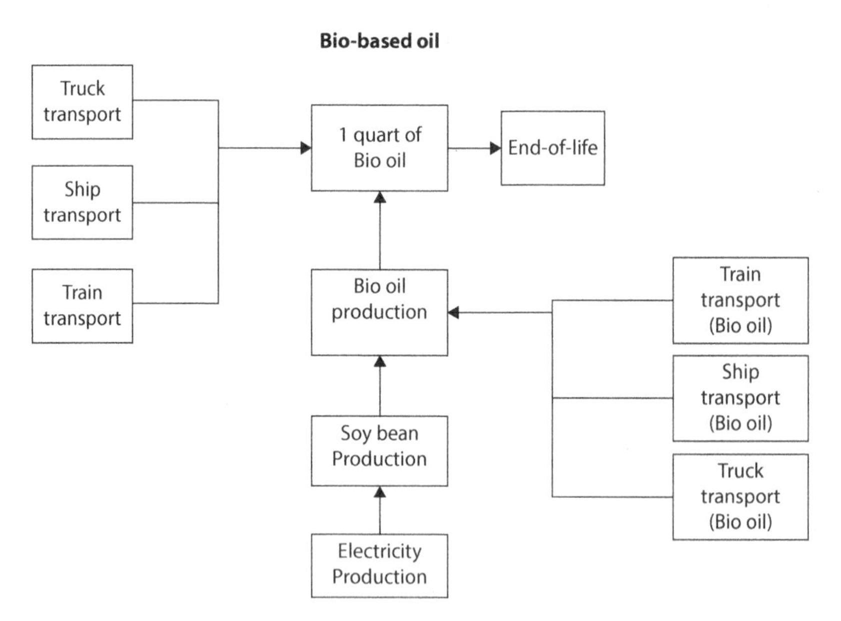

Figure 3 Flow diagram for production and use of bio-based motor oil.

6. Selecting Impact Categories
   Part of the goal statement is the identification of the impact categories that are to be modeled as part of the assessment, and, as required by the ISO standard, are relevant to the goal and scope of the study.
   a) For the laundry detergent example, list some relevant impact category indicators. What could be considered irrelevant?
   b) A Carbon Footprint study is an example of identifying a single impact that is considered relevant in the application of the LCA. In such cases, only potential contributions to global climate change are inventoried and characterized in impact models. CFs are easier to conduct and interpret than more complete LCAs because they greatly simplify the process. Discuss the benefits and disadvantages of CF and other single issue approaches. How does focusing on one impact category possibly overlook other important impacts?

3

Life Cycle Inventory

Abstract
The life cycle inventory (LCI) of flows to and from the natural environment provides the foundation for conducting a life cycle assessment (LCA). The LCI also provides the basis for the impact assessment and interpretation phases of the LCA, so it is critically important that the inventory be sound, complete, and unbiased. This chapter presents important methodological issues where different modeling choices must be made when creating an LCI, depending on the specific system being studied. These decision points include cut-off rules, co-product allocation schemes, recycling (postconsumer and industrial scrap), system expansion and avoided burden considerations. Other methodological issues raised, due to increasing interest in the environmental community, include water use and carbon tracking. The chapter goes on to discuss data sources and the many LCA software products that are available. Finally, examples of a data collection form and a reporting form are provided.

References from the LCA Handbook

Aims of the Chapter

1. Describe how to compile and quantify inputs and outputs of the unit processes that make up an industrial system (as defined by in the goal and scope definition phase) to create an LCI.
2. Help readers understand the differences in methodological approaches when creating an LCI and how these choices can affect the final results of the assessment.
3. Describe important considerations in collecting and reporting LCI data.
4. Suggest available databases and data sources that can be used to create LCI.
5. Present available LCA software programs, both free and for purchase.

3.1 Introduction

The ISO standard for Life Cycle Assessment (LCA) defines life cycle inventory (LCI) analysis as the "phase of life cycle assessment involving the compilation and quantification of inputs and outputs for a product throughout its life cycle." It is built on the basis of the unit process (Figure 3.1). A unit process is the "smallest element considered in the life cycle inventory analysis for which input and output data are quantified." Examples of unit process are coal mining, steel production, refining of oil, production of furniture, use of a television set, recycling of waste paper, and transport by lorry (box truck). Each of these processes is described in quantitative terms as having inputs and outputs.

The LCI, then, is the compilation of the different unit processes within the system under study (Figure 3.2). Some unit processes are connected with one another in simple upstream-downstream connections, e.g., TV production is upstream connected

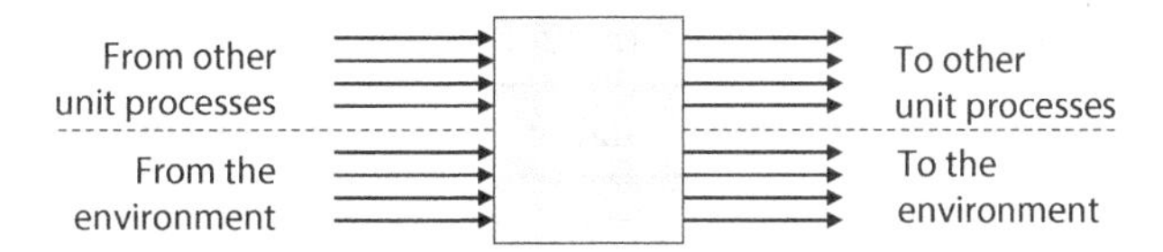

Figure 3.1 General template of a unit process. The process (grey rectangle) is considered as a black box, having inputs (left-hand side) and outputs (right-hand side) from and to other unit process (top lines) and from and to the environment (bottom lines).

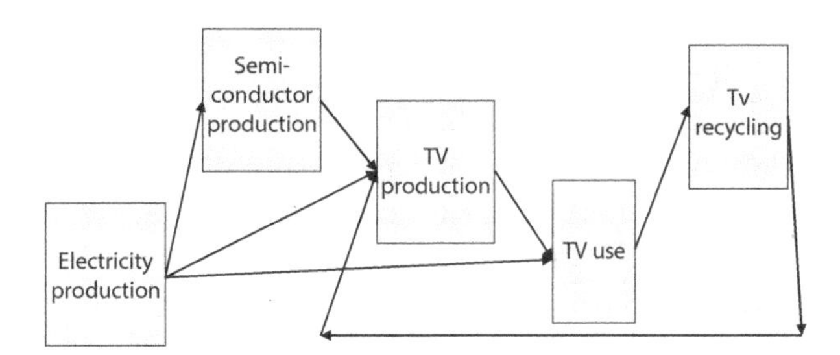

Figure 3.2 Fragment of a simplified flow diagram for an LCA on television (TV) sets. Because the purpose is to show how unit process are connected, only the flows from and to other unit processes are displayed, and flows from and to the environment are hidden. All transport, packaging, etc. has been left out as well.

to semi-conductor production. But there are also more complicated connections, e.g., electricity linking to different parts of the system, and recycling feeding back to production. Flow diagrams are in fact huge webs of interconnected unit processes. In the present era of digital databases, an LCA study can easily be comprised of several thousand unit processes.

3.2 Modeling Inputs and Outputs

In creating LCI, a unit process is treated as a black box that converts a bundle of inputs into a bundle of outputs. Inputs come in several types: products (including components, materials, and services), waste for treatment, and natural resources (including fossils, ores, biotic resources, and land). Outputs come in several types as well: again products (including components, materials, and services), waste for treatment, and residuals to the environment (including pollutants to air, water, and soil, waste heat, and noise).

The LCI phase of an LCA is a quantitative model in which, all unit processes being studied have to be accounted for by specifying the sizes of the inflows and outflows, per unit process. For example a plant producing aluminum may specify their technology in term of inputs and outputs by stating input requirements (e.g., 2 kg aluminum oxide and 20 kWh electricity per kg produced aluminum) and emissions (e.g., 2 g dust per kg produced aluminum). We must translate this into our template for unit processes; see Table 3.1.

Table 3.1 Example of Unit Process Specification for an Aluminum Production Plant.

type of flow	name	amount	unit
inputs from other unit processes	aluminum oxide	2	kg
inputs from other unit processes	electricity	20	kWh
outputs to other unit processes	aluminum	1	kg
outputs to the environment	dust	0.002	kg

In Table 3.1, the unit process data is given per unit of output, here per kg of aluminum. In an LCA, we must next find out how much we need. For instance, the product may need 3 kg of aluminum, not 1 kg. The basic assumption of the LCA model is that technologies are linear. This means that we can scale the data of a unit process by a simple multiplication. In the example, 3 kg of aluminum would require 6 kg of aluminum oxide and 60 kWh of electricity, while it would release 6 g of dust. The assumption of linear technology is an important restriction of LCA; yet it is an important step in making the calculation and data collection feasible.

3.3 Methodology Issues

In this section, several important methodological issues are discussed, including areas where different methodological choices can be made depending on the specific systems being studied, as well as methodological issues that are the focus of increasing interest in the environmental community, such as water use and carbon tracking. In cases where the choice of methodology has a strong influence on the study results and conclusions, the practitioner should justify the reasons for the methodology chosen, and a sensitivity analysis should be conducted to see if an alternative methodological choice produces similar or different results and conclusions.

3.3.1 Cut-Off Rules

Cut-off is a solution to the problem that the system is theoretically infinitely large. To produce a television set (TV), we need machines, and these machines are produced by machines, and these machines in turn need machines, etc. But of course we have an intuitive idea that some very distant upstream processes will be quite unimportant to the study. This means that we will cut-off certain inputs: although we know that something is needed, and we sometimes even know how much is needed, we do not go to the trouble of specifying how these inputs are produced. It turns out to be difficult to specify reliable criteria when cut-off is allowed, or to estimate how large the error is when a cut-off is made. Criteria on the basis of negligible contribution to mass or cost (e.g., smaller than 1%) often work pretty well, but occasionally have been shown to yield large errors. Alternatively, estimates of missing parts by means of similar processes (e.g., estimating the production of a freezer by using production data for a refrigerator), or by economic input-output tables may be helpful. Another approach is to conduct a difference analysis, leaving out the activity in the system that functions the same. For example, in comparing a cathode ray tube (CRT) to liquid crystal display (LCD) TV we may leave out the broadcasting processes.

Criteria for excluding components or materials are defined at the outset of the project but may change based on limitations encountered as the study is conducted. Cut-off rules are typically expressed in terms of mass, for example, "the study will account for at least 95% of the total mass of inputs, and no input shall be excluded that individually contributes 1% or more of the mass." Ideally, a life cycle study would account for all life cycle steps and 100% of the content of product, modeled using data for the actual materials and processes. Practically, data are often not available for some processes or materials or cannot be gathered within the time and budget constraints for the study. This is often true in comparative analyses where the organization sponsoring the study can provide detailed data on their own system, but alternative systems must be modeled using publicly available data.

Before deciding to exclude materials or processes from the study, it is important to carefully consider the potential effect on study results. Mass contribution is usually the criterion used to identify components for possible exclusion, but a material with a small mass contribution may have significant impacts on energy or environmental impacts. For example, an exterior metal plating a few microns thick may add only a tiny amount to the mass of a product. However, the energy and emissions associated with the production of the metal, or metal emissions from the plating process, may have impacts that are large relative to the mass of material used in the product.

Another example would be a thin exterior coating that is cured in an energy-intensive baking process. Even if data on production of the specific coating material is not available, energy for the baking process should be included because of its contribution to energy impacts.

Sometimes additional research can uncover options for inclusion of components for which data do not initially seem to be available. Take the example of a molded plastic product that has inputs of a large quantity of resin and a much smaller quantity of a material identified as a color compound. Because the color compound represents a small percentage of the total mass of inputs, and data on production of pigments are generally not available, the first inclination may be to exclude the color compound. However, manufacturer literature for the compound indicates that the compound consists of a small amount of pigment blended into the same resin used as the main input to the process. In this situation, the amount of resin in the color compound can be included in the modeling, increasing the total mass of inputs covered in the analysis. Only the very small amount of pigment in the color compound would still be excluded. Information in manufacturer literature and material safety data sheets (MSDS) can often help identify constituents of specialty materials so that at least some of the content of these products can be included or suitable surrogate data can be selected.

Another cut-off decision involves capital goods and infrastructure, i.e., the buildings and equipment used to manufacture the product, or the vehicles used to transport products. In the past, sample calculations suggested that the contributions of infrastructure tend to be small when allocated over the total amount of output over the lifetime of the infrastructure. However, more recent analyses have indicated that infrastructure contributions may be significant for certain industry sectors, e.g., non-fossil electricity generation (Frischknecht *et al.* 2007).

3.3.2 Co-Product Allocation

As shown in Figure 3.3, many processes produce more than one useful output and are considered "multi-functional." This makes it necessary to use some method to divide or partition the process input and output flows among the useful outputs. Because of the many ways this can be done, co-product allocation has given rise to one of the biggest controversies in LCA theory.

Within ISO 14044 (section 4.3.4.2), a preferred order for solving the multi-functionality problem has been designed.

- Avoid allocation where possible,
- Where allocation cannot be avoided, allocate inputs and outputs among useful coproducts in a way that reflects physical relationships,
- Where physical relationship alone cannot be used, allocate based on other relationships.

Two options are given for avoiding allocation. The first option is to further subdivide the given process into subprocesses with inputs and outputs that can be assigned to individual co-products. This approach can be used, for example, when operating data on a manufacturing facility are provided as a "black box," but individual co-products can be traced to separate processes within the facility. In many cases, however, even at the most detailed subprocess level, a single process produces multiple co-products.

The second option is to avoid allocation by expanding the system boundaries. Figure 3.4 shows an example of comparing Product A with Product C. Since Process 1 makes B in addition A, Product C's system is expanded to include Product D which provides the same function as Product B. The comparison is then made between Process 1 and Processes 2+4.

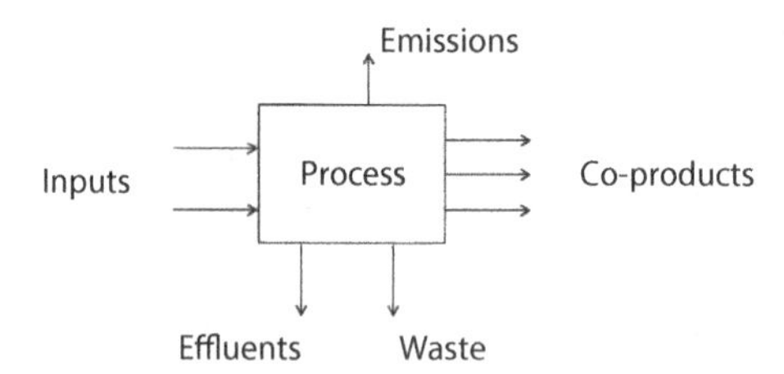

Figure 3.3 Multiple Co-Products from a Unit Process.

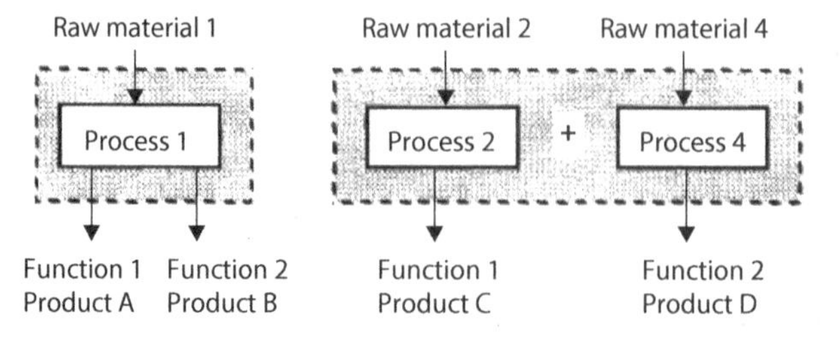

Figure 3.4 System Expansion of Product C, being compared with Product A (Tillman *et al* 1994).

Comparing corn-based ethanol to petroleum-based gasoline as automotive fuels is an example of this application of system expansion. A typical corn mill produces not only corn grain (Product A), but also co-products such as corn oil (Product B). Gasoline (Product C) would then be modeled along with a product that performs the same function as corn oil, such as soybean oil (Product D).

When allocation cannot be avoided, physical relationships such as mass or energy are commonly used as the basis for allocating process flows among useful outputs. The allocation should be related to the function of the products. Using the hydrocracker example, the ethylene, propylene, and other co-products are used as material inputs for production of plastic resins and other petrochemicals rather than as fuels, so burdens can be allocated among the co-products based on their mass rather than on their energy value.

Economic value may also be used as a basis for allocation; however, this approach should be used with caution, since fluctuations in price of co-products can change the results and conclusions of an analysis, even if there has been no change in the physical relationship between the co-products.

This stepwise procedure is a clear comprise, and in practice leaves much freedom in LCA practice potentially resulting in studies giving conflicting results.

After appropriate cut-off and allocation steps, the final inventory results can be calculated. Typically, this is a table with the quantified inputs from and outputs to the environment, for each of the alternative systems considered, expressed in relation to the functional unit. With the present-day software and databases, this inventory table may be a thousand lines or longer. It contains not only the familiar pollutants and resources, such as CO_2, NO_x, and crude oil, but also more exotic items, such as 1-pentanol, cyprodinil, and dolomite.

In scaling unit processes, the web-like nature of the system quickly creates complications, as everything depends upon everything. The calculation of the scaling factors, and with that of the emissions to and extraction from the environment, is greatly simplified by considering the problem as a system of linear equations: one unknown (the scaling factor) for every unit process, and one equation (a balance) for every flow. Thus, solutions may be obtained by matrix algebra. The details of this are not discussed here; see Heijungs and Suh (2002) for a detailed exposition.

The approach mentioned above may fail in a number of cases. We mention two complications:

- For some products, upstream production processes or downstream disposal processes may be difficult to quantify;
- For some unit processes, the balance equations become impossible due to the fact that these processes produce not just one product but several co-products.

The first issue can be solved by the cut-off procedure; the second one by allocation.

The example problem can be demonstrated thus: If a transportation process needs gasoline, the upstream unit process is a refinery that produced not only gasoline, but also diesel, kerosene, heavy oils, and many other products. The direct impacts (from pollutants like CO_2), but also the flows to and from other processes that may lead to impacts (e.g., from oil drilling) may be argued not be attributable to gasoline only, but need to be distributed over gasoline, diesel, and all the other co-products. This is hardly contested, but the debate focuses now on how to do this. To make the issue more concrete, the question at hand can be stated as: How much of the CO_2 from a refinery is allocated to the gasoline? Different schools have provided different arguments, none of which have been completely compelling so far. Some solutions lead to strange results, while other solutions may be very difficult to carry out (e.g., for lack of data or appropriate software). Still others are rejected outright by many experts. To complicate the issue, the problem does not only occur in unit process that produce several co-products, but also in unit processes that treat more than one type of waste, or where waste is recycled into a useable good. It is not even agreed upon if the multi-output case, the multi-input case, and the recycling case must be treated in the same way or not.

3.3.3 Postconsumer Recycling

There are a number of approaches that can be used for modeling the impacts of postconsumer recycling. Commonly used approaches include system expansion, boundaries drawn between successive useful lives of the material, and allocated approaches.

3.3.3.1 System Expansion and Avoided Burden

Besides the ISO-based "expanding the system to include the additional functions," we often see a method that is best described as "subtracting the avoided impacts from additional functions." This is commonly known as the substitution method or the avoided burden method. All process burdens are assigned to the primary product of interest, and credit is given for materials or services that are displaced by the other co-products of the process. System expansion may not be a suitable approach if the process being evaluated is the primary or only commercial route for producing one or more of the co-products, so that there is not a basis for displacement credit. A hydrocracker unit uses inputs of refined oil and natural gas to produce outputs of ethylene, propylene, other hydrocarbons, fuel gas, and heat. An energy credit can be applied for the fuel gas and heat co-products used outside the system boundaries, but the remaining process burdens must be allocated among the material co-products.

For instance, when a waste treatment activity co-produces electricity, the emissions from the regular way of producing the same amount of electricity are subtracted. This method has similarities with that of system expansion, but of course they are not identical. Many LCA studies employing the substitution method claim to be ISO-compliant, even though strictly speaking ISO 14044 does not mention this method, let alone recommend it. That does not necessarily mean that these studies are incorrect, of course. Compliance with ISO is not a sufficient quality guarantee, but also not a necessary one.

For material that is recycled at end of life, the system that produced the material is assigned the burdens for collection of the postconsumer material and reprocessing the recovered material into a form that is ready for its next use. Credit is then given for avoided production of the material that is displaced by the recycled material. The credit is given to the system *producing* the recycled material. The rationale is that because the first system is *supplying* material as a feedstock for future systems, less virgin material must be produced.

3.3.3.2 Cut-Off Method

This approach also avoids the need for allocation. Distinct boundaries are drawn between systems producing and using recycled material. The initial system is assigned all virgin production burdens for the material, and material going to recycling leaves the first system's boundaries at end of life. The user system (second system) burdens begin with collection and reprocessing of postconsumer material.[1]

Since collection and reprocessing burdens are generally much lower than virgin production burdens, this approach tends to favor the system *using* recycled material. The rationale can be expressed as follows: Because the second system is *using* recycled material, demand for virgin material is reduced and less virgin material must be produced.

3.3.3.3 Allocated Burdens

The rationale for the allocated approach is that all product systems using a given quantity of material should share equally in recycling burdens and benefits. This includes the system first using the material in virgin form as well as all subsequent systems using the material after recovery and reprocessing.

The following equation can be used to illustrate the general concept of the allocated approach:

$$V/n + F + U + (n-1)/n \times R + D/n$$

where V = virgin production, F = fabrication, U = use, R = recycling, D = disposal, and n = the total number of useful lives of the material (including virgin use and all subsequent uses until the material is disposed). The recycling allocation factor (n-1) is one less than the total number of uses since there is no recycling preceding the initial use.

In reality, with each recycling cycle there will be collection and reprocessing losses, and subsequent uses of the recycled material are likely to be in products with a mix of virgin and recycled content. To illustrate the general concept, however, we will use the simplified equation, with 100% recovery and recycling after each use.

Allocation Equation Applied to Open Loop Recycling: Open loop recycling describes a system in which a product is recovered at the end of its useful life, and the recovered material is then used in a different type of product system. Typically the second product is disposed after use, or the material may be recovered and reused in a product that has a low recycling rate and therefore a low probability of repeated use cycles. Open loop recycling often applies to materials with properties that degrade with repeated use cycles, for example, paper fibers that become shorter with each repulping and remanufacturing cycle. Open loop recycling also applies to products that have low recycling rates even though the material properties may be suitable for repeated use cycles. In open loop recycling, the total number of useful lives of the material "n" is a small number. For example, if material is used in a virgin product, recovered and recycled into a second product, and the second product is disposed at end of life, then n=2, and each

---

[1] This method is outlined in EPA/600/R-92/245 **Life-Cycle Assessment: Inventory Guidelines and Principles**, where it is identified as recycling allocation method 2.

system that uses the material is allocated half the virgin production burdens, recycling burdens, and disposal burdens.

Allocation Equation Applied to Closed Loop Recycling: Closed loop recycling occurs when material is used in a product, the product is recovered at end of life, and the recovered material goes back into the same type of product, so that there are repeated recovery and reuse cycles. In order to get a large number of use cycles out of the material, the material properties must hold up through repeated use cycles (e.g., glass, metals). As the total number of uses "n" increases, the virgin production burdens and disposal burdens allocated to each use become smaller, and (n-1)/n (the allocation factor for recycling burdens) approaches 1.

There are limitations to the allocated recycling approach. This method requires assumptions about the total number of lifetime uses of the material. For a given product application, it is only possible to state with certainty whether incoming material is virgin material or postconsumer material. If the material enters the system as postconsumer material, it has had at least one previous life, but it is not possible to determine

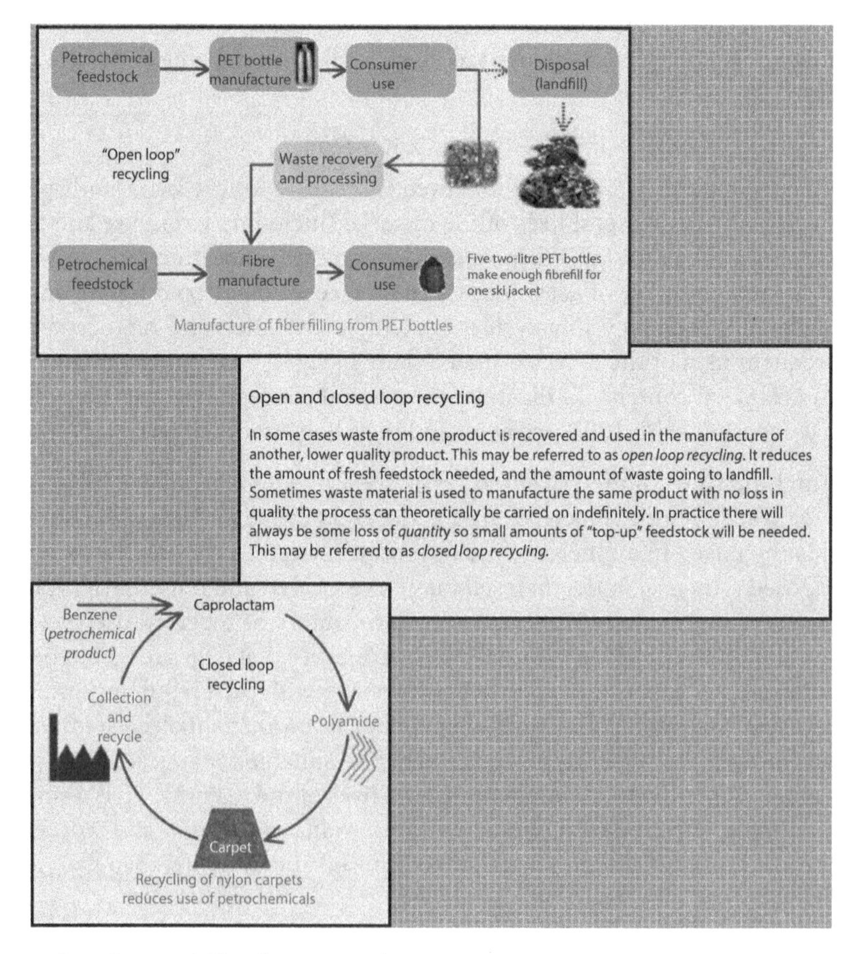

Figure 3.5 Open Loop and Closed Loop Recycling. http://www.greener-industry.org.uk/pages/recycling/2recyc_understand.htm.

the total number of previous uses. Similarly, if the current product is known to be recycled at end of life, the material will have at least one subsequent use, but the fate of the material after the next use is uncertain.

For durable products that have very long useful lives, there is additional uncertainty about future recycling. If the product is in use for many years, recycling rates and technologies at the end of the product's useful life may be quite different from recycling rates and practices at the time the product was manufactured.

Allocation calculations can become very complicated when adjusting for reprocessing losses and sequential useful lives that have different mixes of virgin and postconsumer content and different postconsumer recovery rates. Because the system expansion and cut-off approaches focus on the material's use in the current product system and the material's next use, the calculations for these methods are more straightforward and require fewer assumptions about prior and future uses of the material.

For recycling to be sustainable, there must be a balance between recycled material supply and demand. If the supply of recycled material used by the system is currently fully utilized (e.g., secondary aluminum), then the recycling rate for a product must be equal to or greater than its recycled content in order to be sustainable. If a system uses more of a fully utilized recycled material than it produces (recycled content > recycling rate), then it creates a net deficit in the recycled material supply that has to be made up with virgin material. If the system's recycling rate is higher than its recycled content, then it is a net producer of recycled material, and a credit can be applied for the virgin material that is displaced by the surplus recovered material.

3.3.4 Converting Scrap

Scrap that is generated during material converting processes is referred to as postindustrial or preconsumer scrap. Unlike postconsumer scrap, preconsumer material has not had a previous useful life in a product, so there is no previous life to allocate virgin material burdens. However, the material is often degraded to some extent during the converting process. For example, the material may have been coated, had colorants added, or been glued or laminated to other materials. The industrial scrap material usually requires some degree of reprocessing before it can be used to produce a useful product.

If the converting scrap is utilized internally at the same facility in the same process that produced the scrap, then this internal recycling simply reduces the net amount of virgin inputs required per unit of product output, and no allocations are needed. The process burdens for manufacturing the primary product should include the added burdens for any reprocessing of internal scrap before it is returned to the process (e.g., regrinding of plastic molding scrap before it is put back into an extruder).

If the scrap is used outside the boundaries of the system that produces the scrap, then there are different approaches that can be used to allocate the virgin material burdens associated with the scrap material. It is important to distinguish between the burdens associated with production of the *material content* of the scrap and the burdens associated with the *process* that generates the scrap. The burdens for the converting *process* that generates the scrap should be assigned to the primary product, since the converting process adds no value to the scrap material and usually reduces its value. For example, the processes of applying coating to carton board and cutting it into carton blanks are done for the purpose of producing a finished carton blank, so the environmental burdens for these processes should be assigned to the carton blanks. The coating on the trim scrap generated from the converting process reduces the value of the scrap, because the coated scrap must be repulped and the fiber separated from the coatings before the fiber can be used to make a useful product.

When deciding whether it is appropriate to apply credits for using postindustrial scrap, one should consider the current utilization of this type of scrap. If the scrap is currently being fully utilized (e.g., kraft clippings from box manufacture), use of the scrap diverts the material from some other use rather than diverting it from disposal. Material production burdens can be allocated between the primary product and the scrap based on the mass of input material that ends up in the finished product and the mass of scrap that leaves the system boundaries as input to another product.

However, if the scrap requires additional reprocessing, as in the example of coated carton board scrap, then the scrap material will be used only if the extra reprocessing steps can be economically justified. In this situation, virgin material burdens may be allocated between the primary product and the converting scrap based on the relative quantities of material that end up in the primary product and in the converting scrap, and the relative economic value of the virgin material and the devalued scrap.

3.3.5 Water Use

In recent years water use, in particular freshwater use, has become an area of increased interest in life cycle inventories and assessments. However, a large body of life cycle data has been developed over the years without gathering accompanying water use data, making it necessary to add water use to data sets where it is missing.

A similar situation existed when global warming became recognized as an important environmental issue. Prior to that time, carbon dioxide emissions had not been tracked databases, used later for LCI, because there had been no environmental reporting requirements for carbon dioxide. Adding carbon dioxide emissions to LCI databases was relatively easy, however, since carbon dioxide emissions can be estimated based on the carbon content of material. Water use, on the other hand, is more difficult to characterize and quantify, since there are many forms of water use and different types and sources of water that can be utilized.

Water use is a broad term that can include any form of use that makes water permanently or temporarily unavailable for use by another system. Water use can be generally classified as in-stream use or off-stream use. Off-stream use involves withdrawal from a water source, while in-stream uses do not. Examples of in-stream use are hydroelectric generation, water transport, fisheries, or recreational uses such as boating.

Off-stream uses involving water withdrawals can be further classified as degradative or consumptive (Koehler 2008). Degradative use returns the water to the same watershed from which it was withdrawn, but with changes in quality (e.g., addition of contaminants, temperature changes). Consumptive use refers to water that is withdrawn from one source and returned to a different water body or watershed (e.g., depleting the initial water source but adding to the receiving water), or water that is withdrawn and not directly returned to a receiving body (such as water embodied in a product or evaporated in a cooling tower or drying operation). Consumptive use of water can be a very significant environmental concern, particularly in areas where fresh water is scarce. There is less focus on consumptive use of saltwater.

Protocols are being established for categorizing and reporting the various types of water use. There is an active UNEP/SETAC working group on water use and consumption within LCA. As of June 2011, the International Organization for Standardization initiated a working draft of a new standard, ISO 14046 Life cycle assessment -- Water footprint -- Requirements and guidelines, to provide internationally harmonized metrics for water footprints.

3.3.6 Carbon Tracking Considerations

Carbon is tracked as two different forms in life cycle methodology: biogenic carbon and fossil carbon. Biogenic carbon is carbon that is removed from the atmosphere and incorporated into the physical mass of a plant or organism. The carbon remains embodied in the biomass-derived product until its useful life when some or all of the carbon is either somehow sequestered (e.g., if the biomass-derived product is landfilled and some of the material does not decompose) or returned to the atmosphere due to decomposition or combustion. Therefore, biogenic carbon returns to the atmosphere in the same form as which it was removed, resulting in no net increase in atmospheric carbon dioxide within the time frame of natural biogenic carbon cycling. These biomass carbon dioxide emissions are considered to be "carbon neutral."

On the other hand, emissions associated with the combustion of fossil fuels or fossil-derived materials are treated as net contributors to atmospheric carbon dioxide levels Although the carbon originated as biogenic, uptake occurred millions of years ago and the carbon remain stored within the earth until released through human intervention. No carbon storage credit is given when materials such as fossil fuel-derived plastics are landfilled (EPA 2006).

There are additional considerations regarding decomposition of landfilled biomass products. If a biomass-derived product decomposes aerobically, the carbon dioxide released is considered carbon neutral. However, if the biomass decomposes anaerobically, both carbon dioxide and methane will be produced. For either type of decomposition, the carbon dioxide produced is considered carbon neutral, but the methane is not. Since human intervention in the biomass carbon cycle is responsible for some of the atmospheric carbon returning to the atmosphere as methane, with a higher global warming potential than the carbon dioxide initially taken up by the biomass, the methane releases are not considered carbon neutral.

Similar carbon tracking issues apply to waste-to-energy combustion of materials. Carbon dioxide from combustion of biomass-derived material is considered carbon neutral, while carbon dioxide from the combustion of materials derived from fossil fuels is considered as a net contribution to global warming potential. Regardless of whether the carbon in the combusted material is biogenic or fossil carbon, credit should be given for the energy and emissions displaced by energy recovered from combustion of the material.

End of life carbon tracking calculations can become quite complicated when considering the potential mix of fates of biomass products and the time frame over which releases occur. As noted previously, biomass-derived products may decompose in landfills, but this is subject to landfill conditions (e.g., temperature, moisture, presence of microbes). It may take many years for the decomposition to occur, and the decomposition may never completely convert all the carbon content to carbon dioxide and methane. Samples of newspaper and other bio-derived products excavated from actual landfills have shown very little degradation [11]. Landfill simulation studies have also indicated that the lignin content of products derived from woody biomass tends not to decompose [12]. Biomass decomposition can also be inhibited by moisture-resistant coatings, fillers and additives, or sandwiching biomass layers between layers of foil or plastic.

If landfilled biomass does decompose anaerobically, there are different possible fates for the methane that is generated. If the methane escapes into the atmosphere uncaptured and untreated, it results in additional global warming potential. If the methane is captured and flared (with or without energy recovery) or oxidizes as it travels through the landfill cover, then the carbon content returns to the atmosphere as carbon-neutral CO_2. If the captured methane is burned with energy recovery, then the useful energy recovered can displace natural gas or electricity consumption, and credit should be given for the displaced energy and emissions.

3.4 Data Uncertainty and Sensitivity Analysis

Many LCA studies continue to be conducted without uncertainty or sensitivity analysis, even though software increasingly facilitates this. Ross *et al.* (2002) surveyed 30 studies and observed that 47% mentioned uncertainty, 7% performed qualitative analysis and 3% performed quantitative uncertainty analysis.

There is of course a psychological argument that a contractor pays for finding out something, not for increasing the doubt. And as many LCA practitioners spend several months on collecting data, it is never a nice thing to waste this effort in a last-minute uncertainty analysis. But decision-making obviously means also taking into account the limits of knowledge. Moreover, as discussed before, a proper analysis of uncertainties and sensitivities helps to prioritize the steps earlier on in the framework: collecting data, setting boundaries, making choices.

The iterative nature of the ISO framework shows up in this context. Whenever the uncertainties are too high, we may go back to collect better data. Whenever sensitivity analysis shows that some decisions are crucial, we may go back and do a more refined analysis. Uncertainty analysis and sensitivity analysis are discussed further in Chapter 6 on Interpretation and Reporting. Interpretation aims to prepare for a balanced decision, while helping improve the LCA.

3.5 Databases and Data Sources

In many instances, creating an LCI begins with the collection of raw data which are data extracted from various sources, such as bookkeeping of a plant, national statistics, technical journals, etc., but not yet related to the process for which the dataset is being developed. Typically, a number of sources are needed to be called upon to collect a sufficient amount of data. Other examples of data sources that may be drawn from or utilized include the following:

Table 3 Types of Data Sources used to Create Life Cycle inventory

- Individual process- and facility-specific: data from a particular operation within a given facility that are not combined in any way.
- Composite: data from the same operation or activity combined across locations.
- Aggregated: data combining more than one process operation.
- Industry-average: data derived from a representative sample of locations and believed to statistically describe the typical operation across technologies.
- Generic: data whose representativeness may be unknown but which are qualitatively descriptive of a process or technology.

Data can be classified by how they are created:

- Site-specific (directly measured or sampled)
- Modeled, calculated or estimated
- Non-site specific (i.e., surrogate data)
- Non-LCI data (i.e., data not originally intended for use in an LCI
- Vendor data

Data sources are either primary or secondary:

1. Primary data come directly from the source, including:

- Interviews,
- Questionnaires or surveys,
- Bookkeeping or enterprise resource planning (ERP) system,
- Data collection tools (online, offline), and
- On-site measurements.

2. Secondary data come from reports found in:

- Databases,
- Statistics, and
- Open literature.

- Meter readings from equipment
- Equipment operating logs/journals
- Industry data reports, databases, or consultants
- Laboratory test results
- Government documents, reports, databases, and clearinghouses
- Other publicly available databases or clearinghouses
- Journals, papers, books, and patents
- Reference books
- Trade associations
- Related/previous life cycle inventory studies
- Equipment and process specifications
- Best engineering judgment

Once raw data are collected, following a pre-determined data collection approach (see section 3.6 Collecting LCI Data), unit process datasets can be created by defining mathematical relationships between the raw data and various flows associated with the dataset in a defined reference flow. Data modeling requirements, with desired quality attributes and adequate documentation, are specified to accurately transform raw data into unit process datasets, and incorporate proper review and documentation to address verification and transparency issues (Consoli, Allen *et al.* 1993; Curran 2011). Therefore, understanding how data flow from raw data providers to LCI data users (shown in Figure 3.6) is important because data move from the raw state to and through datasets and databases.

For emissions, it is preferable to gather data on process emissions released to the environment after any on-site controls or treatment have been applied. If the reported

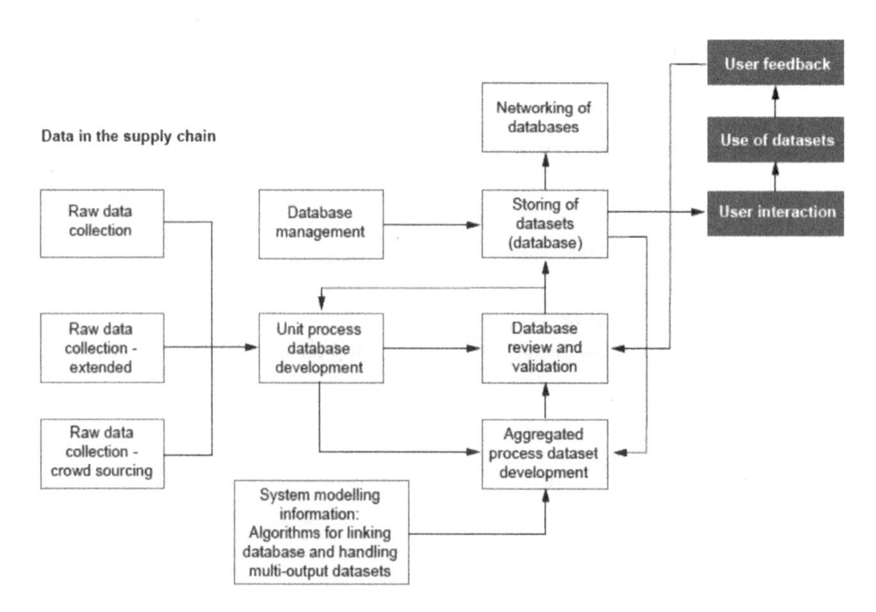

Figure 3.6 Flow of Data from Raw Data through to the LCI Data User with Feedback Loops (UNEP/SETAC 2011).

emissions include fuel combustion emissions, it is important that this be noted, so that fuel-related emissions are not double-counted later when constructing the life cycle inventory model and linking to data sets for the reported process fuels.

Emissions should be speciated to the extent possible in order to facilitate subsequent impact assessment. Instead of reporting a group of emissions as volatile organic compounds (VOC), the chemical composition of the emissions should be reported. Different isomers of a chemical can have different human health and ecotoxicity impacts (e.g., ortho- and para-xylene), so it desirable to speciate emissions as precisely as possible.

Unit process datasets are the basis of every LCI database and the foundation of all LCA applications. A unit process dataset is obtained as a result of quantifying inputs and outputs in relation to a quantitative reference flow from a specific process. These inputs and outputs are generated from mathematical relationships based on raw data that have not previously been related to the same reference flow. An aggregated process dataset is obtained from a collection of similar unit process or other aggregated datasets. Most often, datasets are aggregated to protect business-sensitive, competition-sensitive, or proprietary information, including trade secrets, patented processes, process information used to easily derive costs, etc.

"Unit process" is defined as "smallest element considered in the life cycle inventory analysis" in ISO 14040 (ISO 2006). Unit process datasets are usually distinguished from aggregated process datasets. However, when used in creating an LCI, an aggregated process dataset may be considered as representing a unit process.

The required level of aggregated data should be specified (as guided by the study's goal), for example, whether data are representative of one process or of several processes. Figure 3.7 depicts the possible variations to aggregate processes (steps 2 through 11). Step 1 indicates no aggregation (a single process); step 12 is the complete cradle-to-grave LCI, the ultimate form of aggregation.

3.5.1 Private Industrial Data

Complete and thorough inventories often require using proprietary data that are provided by either the manufacturer of the product, upstream suppliers, or vendors, or the LCA practitioner performing the study. Confidentiality issues are not relevant for life-cycle inventories conducted by companies using their own facility data for internal purposes. However, the use of proprietary data is a critical issue in inventories conducted for external use and whenever facility-specific data are obtained from external suppliers for internal studies. Consequently, current studies often contain insufficient source and documentation data to permit technically sound external review.

Lack of technically sound data adversely affects the credibility of both the life-cycle inventories and the method for performing them. An individual company's trade secrets and competitive technologies must be protected. When collecting data (and later when reporting the results), the protection of confidential business information should be weighed against the need for a full and detailed analysis or disclosure of information. Some form of selective confidentiality agreements for entities performing life-cycle inventories, as well as formalization of peer review procedures, is often necessary for inventories that will be used in a public forum. Thus, industry data may need to

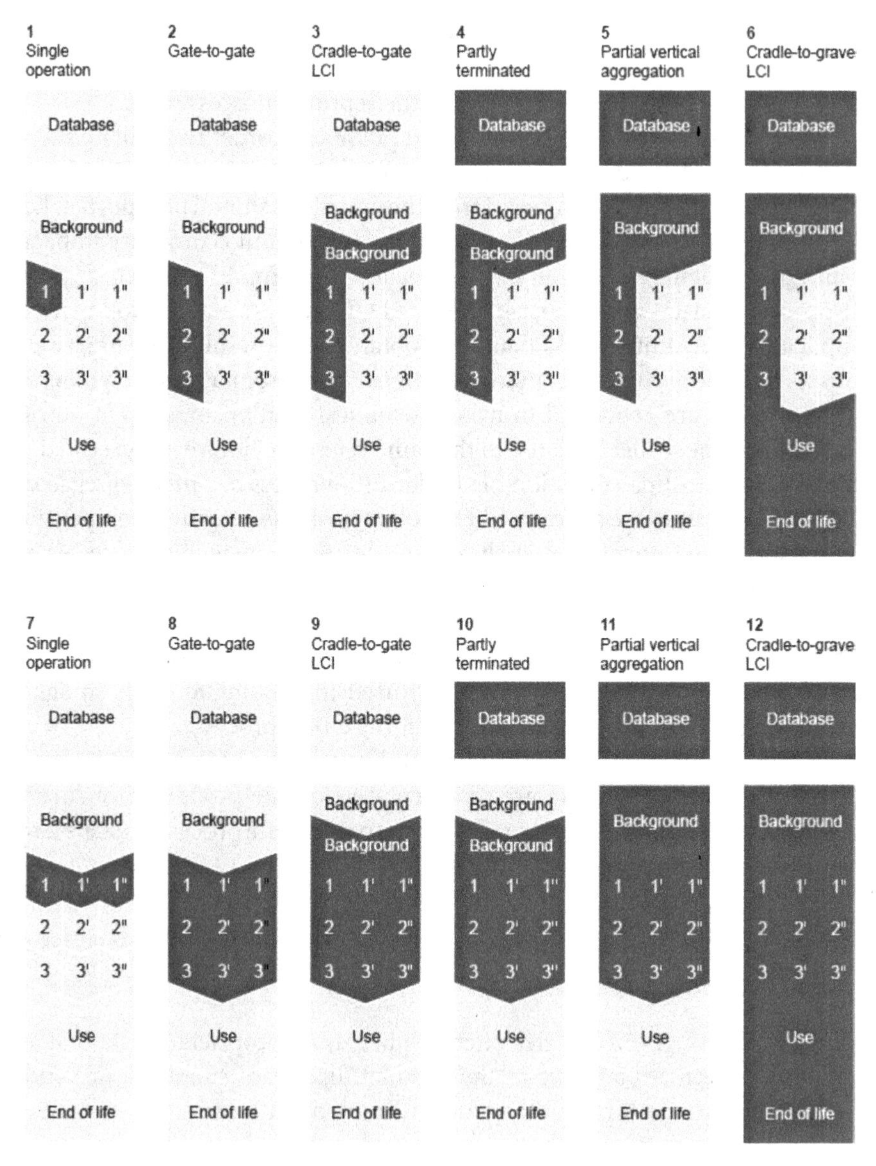

Figure 3.7 Unit process datasets within databases can be aggregated multiple ways, including various combinations of horizontal and vertical aggregation (UNEP/SETAC 2011).

undergo intermediate confidential review prior to becoming an aggregated data source for a document that is to be publicly released.

Examples of private industry data sources include independent or internal reports, periodic measurements, accounting or engineering reports or data sets, specific measurements, and machine specifications. One particular issue of interest in considering industrial sources, whether or not a formal public data set is established, is the influence of industry and related technical associations to enhance the accuracy, representativeness, and up-to-datedness of the collected data. Such associations may be willing, without providing specific data, to confirm that certain data (about which their members are knowledgeable) are realistic.

3.5.2 Public Industrial Data

Technical books, reports, conference papers, and articles published in technical journals are a good source for information and data on industrial processes and activities. Most are publicly available, although data presented in these sources are often older, and they can be either too specific or not specific enough. Many of these documents give theoretical data rather than real data for processes. Such data may not be representative of actual processes or may deal with new technologies not commercially tested. In using the technical data sources in the following list, the analyst should consider the date, specificity, and relevancy of the data:

- Kirk-Othmer's *Encyclopedia of Chemical Technology*
- Periodical technical journals such as *Journal of the Water Environment Federation*
- Proceedings from technical conferences
- Textbooks on various applied sciences

Frequently, the end user will not be able to supply specific information on inputs and outputs. However, the end user can provide data on user practices from which inputs and outputs can be derived. Generally, the end user can be the source of related information from which the energy, materials, and pollutant release inventory can be derived. (An exception would be an institutional or commercial end user who may have some information on energy consumption or water effluents.) Market research firms can often provide qualitative and quantitative usage and customer preference data without the analyst having to perform independent market surveys (EPA 2006).

3.5.3 Dedicated LCI databases

Since the early 1990s, LCA databases have proliferated in response to the growing demand for life cycle information. These data sources have mostly emanated from Northeast Asia, North America, and Western Europe. In a global economy, of course, products and services are sourced from many countries. LCA databases mainly provide life cycle inventory (LCI) data, although characterization factors associated with life cycle impact assessment methods are often included as well.

An increasing number of LCI/LCA databases[2] have been created in response to growing demand. While national databases, typically, a number of sources are needed to collect a sufficient amount of data. Including commercial off-the-shelf software tool, such as GaBi or SimaPro, private industrial data and publicly available data (see section 3.12 LCA Software).

Once raw data are collected, following a pre-determined data collection approach, unit process datasets can be created by defining mathematical relationships between the raw data and various flows associated with the dataset in a defined reference flow. Data modeling requirements, with desired quality attributes and adequate documentation,

[2] Databases containing only inventory data only are sometimes referred to as LCA databases, although the name "LCI Database" is technically correct.

are specified to accurately transform raw data into unit process datasets, and incorporate proper review and documentation to address verification and transparency issues (Consoli, Allen *et al.* 1993, Curran 2011).

Table 3.2 lists national inventory databases and databases created by industrial organizations. The appendix to this chapter presents Life Cycle Inventory Data Sources, prepared by Joyce Cooper at the University of Washington, Seattle, Washington, USA (http://faculty.washington.edu/cooperjs/Definitions/inventory squared.htm).

3.5.4 Non-LCI Data

Government documents and databases provide data on broad categories of processes and are publicly available. Most government documents are published on a periodic basis, e.g., annually, biennially, or every four years. However, the data published within them tend to be at least several years old. Furthermore, the data found in these documents may be less specific and less accurate than industry data for specific facilities or groups of facilities. However, depending on the purpose of the study and the specific data objectives, these limitations may not be critical. All studies should note the age of the data used.

Government databases include both non-bibliographic types where the data items themselves are contained in the database and bibliographic types that consist of references where data may be found. In a study conducted for the US Environmental Protection Agency, Boguski identified twelve data sources that have utility in developing LCA data sets for the US (Boguski 2000)[3]. The data sources Boguski evaluated included the following:

- Aerometric Information Retrieval System (AIRS) – US Environmental Protection Agency
- Permit Compliance System (PCS) - US Environmental Protection Agency
- Biennial Reporting System (BRS) - US Environmental Protection Agency
- Toxics Release Inventory (TRI) Database - US Environmental Protection Agency
- Industrial Assessment Center Database (IAC) – US Department of Energy
- Manufacturing Energy Consumption Survey (MECS) – US Energy Information Administration
- Reasonably Available Control Technology/Best Available Control Technology/Lowest Achievable Emissions Rate (RACT/BACT/LAER) Clearinghouse (RBLC) - US Environmental Protection Agency
- Compilation of Air Pollutant Emission Factors AP-42, Volume I: Stationary Point and Area Sources - US Environmental Protection Agency

[3] Other useful U.S. government sources include: U.S. Department of Commerce's Census of Manufacturers, U.S. Bureau of Mines' Census of Mineral Industries, and the U.S. Department of Energy's Monthly Energy Review.

Table 3.2 Available National Life Cycle Inventory Databases and Industry Organizations' Databases (updated Curran and Notten 2006).

Name	Website	Availability	Language	Data focus (if any)	Number of datasets	Geographic origin
Australian Life Cycle Inventory Data Project	http://www.auslci.com.au/	Free	English		>100	Australia
BUWAL 250	http://svi-verpackung.ch/de/Services/1&Publikationen/	Fee or included with SimaPro	German, French	Packaging materials		Switzerland
Canadian Raw Materials Database	http://crmd.uwaterloo.ca/	Free with registration	English, French	Aluminum, glass, plastics, steel, and wood	17	Canada
ecoinvent	www.ecoinvent.ch	License fee	English		>4000	Global/ Europe/ Switzerland
EDIP	www.lca-center.dk	License fee	Danish		>100	Denmark
German Network on Life Cycle Inventory Data	www.lci-network.de	On-going	German, English			Germany
Japan National LCA Project	http://www.jemai.or.jp/english/lca/project.cfm	Fee	Japanese		>600	Japan

(*Continued*)

Table 3.2 (Cont.)

Name	Website	Availability	Language	Data focus (if any)	Number of datasets	Geographic origin
Korean LCI	http://www.kncpc.re.kr	On-going	Korean, English	Energy, chemicals, metal, paper, rubber, polymers, electronic/electric, construction, production process, delivery, disposal, and utility	158	Korea
LCA Food	www.lcafood.dk	Free	Danish, English	Food products and processes		Denmark
SPINE@CPM	http://cpmdatabase.cpm.chalmers.se/	Free	English	-	>700	Global
Swiss Agricultural Life Cycle Assessment Database (SALCA)	http://www.agroscope.admin.ch/oekobilanzen/	Available through ecoinvent or with project cooperation	German, English, French, Italian	Agriculture	>700	Switzerland
Thailand National LCI Database	http://www.thailcidatabase.net/		Thai, English			Thailand
LCA Digital Commons	www.lcacommons.gov	Free with contact	English	Agriculture	>13,000	USA
US LCI Database	https://www.lcacommons.gov/nrel/search	Free with contact	English		>3,000	USA

Industry Organization	Website	Availability	Language	Product Group or Sector	Geographic Coverage
American Plastics Council (APC)	Available from US LCI Database (nrel.gov/lci)	Free	English	Polymers	America
EDP-Norway	www.epd-norge.no	Free	Norwegian, English	Norwegian business (several sectors)	Norway and Europe
European Aluminium Association (EAA)	www.aluminium.org	Free	English	Aluminium production	Europe
European Copper Institute (ECI)	www.copper-life-cycle.org	Free with contact	English	Copper tubes, sheets and wire	Europe
European Federation of Corrugated Board Manufacturers (FEFCO)	www.fefco.org	Free	English	Corrugated Board	Europe
International Iron and Steel Institute (IISI)	www.worldsteel.org	Free with contact	English	Steel	Global
International Zinc Association	info@iza.com	Available to LCA practitioners on request	English	Zinc	Global
ISSF International Stainless steel Forum (ISSF)	www.worldstainless.org/	Free with contact	English, Chinese, Japanese	Stainless steel	Global
KCL (EcoData)	http://www.kcl.fi/	Fee	English	Pulp and paper	Finnish/ Nordic
Nickel Institute	http://www.nickelinstitute.org	Free with contact	English	Nickel	Global

(Continued)

Table 3.2 (Cont.)

Industry Organization	**Website**	**Availability**	**Language**	**Product Group or Sector**	**Geographic Coverage**
PlasticsEurope (formerly APME)	www.plasticseurope.org	Free	English	Plastics	Europe
Volvo EPDs	http://www.volvotrucks.com/dealers-vtc/en-gb/VTBC-EastAnglia/aboutus/environment/environmental_product_declaration	Free	English	Trucks and busses	Europe
World Steel Carbon Footprint	http://www.worldautosteel.org/Environment/Life-Cycle-Assessment/worldsteel-releases-datasets-to-help-lower-carbon-footprint.aspx	Free by request	English	Steel products	Global

- Compilation of Air Pollutant Emission Factors AP-42, Volume II: Mobile Sources - US Environmental Protection Agency
- Locating and Estimating Air Emissions from Sources (A series of L&E documents) - US Environmental Protection Agency
- Factor Information Retrieval (FIRE) - US Environmental Protection Agency
- Sector Notebooks - US Environmental Protection Agency

It is possible to extract meaningful information from public databases for use in LCA studies. It is also possible to develop LCI data sets for some products by using information from public databases. Public databases have several advantages. They are accessible to anyone who wishes to check LCI results. They typically include many more of the specific emissions from industrial facilities than are included in most private LCI databases. They include data directly from U.S. facilities. There is no need to try to convert European data to U.S. conditions. However, there are disadvantages to using public databases. The organization and presentation of data in public databases often makes it difficult to express the values from the various databases in terms that are generally useful for LCA. One challenge is being able to link energy and emission values to production. For example, the MECS database reports annual energy use for industry groups. Likewise, AIRS, TRI, and BRS report annual emissions. The PCS database reports monthly monitoring values, which may be averaged to obtain annual emission estimates. None of these databases ties energy use or emissions to production.

Production information is difficult to obtain. Production on a facility level is usually considered confidential information and is not usually published. The United States Census Bureau reports production, in mass units by SIC code, for only a few industry groups. In addition, facilities are not reported using unique identifiers, leading to difficulty in linking data sources with production rates (for example, when a facility is sold it is reported multiple times under different ownership names). A method for grouping facility data into logical industry groupings and linking the grouped data to grouped production values would benefit the LCA community and still provide confidentiality to industry.

Another challenge to using public databases is the difficulty in aggregating facility data for the precise group of facilities for which one may want data. Aggregating by SIC code generally groups facilities into too broad a category to be useful for LCA. An example is SIC code 3312, Steel Works, Blast Furnaces (including Coke Ovens), and Rolling Establishments. If one is interested in facilities that manufacture only steel tubing, SIC code 3312 is too broad of a category to provide useful information. The ability to aggregate on some smaller segment of industry would be extremely useful. Ideally, the names and addresses of facilities that produce the material or product of interest are known and the researcher can simply collect data for those facilities. This may be the case for LCA studies with a very narrow scope, but is generally not the case for LCA studies that include commodity products.

3.6 Collecting LCI Data

In creating a life cycle inventory, the key information that must be gathered and documented include the following:

- Reporting basis, time period, geographic location, and other useful information for characterizing the process or reporting facility, such as its age, share of total company or industry production, etc.
- Types and quantities of material inputs
- Incoming transportation of material inputs (reported by transportation mode and distance)
- Types and quantities of useful outputs (product of interest, co-products, recycled scrap, recovered heat, etc.)
- Types and quantities of water inputs (emphasis on consumptive use of fresh water)
- Types and quantities of solid wastes and disposition of each type of waste (e.g., landfilled, burned, burned with energy recovery, land applied, etc.)
- Types and quantities of emissions to air
- Types and quantities of emissions to water
- Types and quantities of materials used to package outgoing product

Figure 3.8 presents an example form that can be used for collecting life cycle inventory data. More examples can be found in the ISO 14044 Annex A: Examples of data collection sheets.

3.7 Reporting Life Cycle Inventory

Often, a different type of spreadsheet, or other template, is needed for collecting data and for reporting the results. Reporting formats vary but typically look like the example in Figure 3.9. Individual spreadsheets are created for each process within the life cycle. This spreadsheet reflects the summary of the underlying spreadsheets.

- Process Name and Identifier
- Reference Flow
- Process description
- Name of the Preparer may be included.
- Basis of calculation.
- Inputs.
- Outputs for Air, Water, Solid Waste, Fuel and Water Consumption.

DATA COLLECTION FORM

Company name ______ Contact telephone ______
Contact name ______ Contact e-mail ______
Facility location ______

TIME COVERAGE Month Year
Start ______ ______
End ______ ______

COMPANY COVERAGE
Percent of company output of this material covered by this facility ______

TECHNOLOGY REPRESENTED (brief description)
______
______

INDUSTRY COVERAGE
Percent of industry output of this material covered by this facility ______

REPORTING BASIS (basis on which all data in the form are reported; e.g. 1000 lb of output, specified number of product units, etc.)
Product Output Basis (with units) ______

MATERIAL INPUTS TO PROCESS
List all raw material inputs to the process, normalized to the product output basis specified above.
Include incoming transport information (miles and mode: truck, rail, ship, etc.)

Material name	Amount	Units (lb, kg, etc.)	Distance	Mode	Distance	Mode

Add lines as needed

USEFUL OUTPUTS	Name	Amount	Units (lb, kg, etc.)
Primary product			
Coproducts			
Recovered energy/heat			
Recycled scrap			

Add lines as needed

WATER USAGE	Source (river, lake, ocean, etc.)	Amount	Units (gal, liters, etc.)
Process water			
Intake			
Output			
Cooling water			
Intake			
Output			

ENERGY USE (direct process energy and fuels used to produce process steam)

	Amount	Units (kWh, gal, liters, etc.)
Electricity (grid)		
Electricity (generated onsite)		
Natural gas		
Distillate oil		
Residual oil		
Coal		

Fuel for onsite generation ______

Add lines as needed

SOLID WASTES
Scrap that is recycled on or offsite should be reported in USEFUL OUTPUTS section

Type of waste	Amount	Units (lb, kg, etc.)	Disposition (landfill, burned, burned with energy recovery, etc.)
Process wastes			
Scrap not recycled			
Wastes collected in emission control devices			
Packaging wastes not recycled			

Add lines as needed

Figure 3.8 Life Cycle Inventory Data Collection Form.

EMISSIONS

Report amount of each substance released to environment AFTER on-site emission controls or wastewater treatment.

Emissions to Air

Name	Amount	Units (lb, kg, etc.)	Indicate emissions included: Process	Fuel combustion

Add lines as needed

Emissions to Water

Name	Amount	Units (lb, kg, etc.)	Indicate emissions included: Process	Fuel combustion	Receiving body of water (river, lake, municipal sewer, etc.)

Add lines as needed

PACKAGING FOR OUTGOING SHIPMENT

	Quantity	Units of product (items, lb, kg, etc.) per shipping unit
Product shipped in units of		

List material used per shipping unit of product (corrugated boxes, plastic film sleeves, etc.)

Packaging Material	Amount	Units (lb, kg, etc.)

DATA COLLECTION METHOD AND SOURCES

For each flow, check the appropriate box that describes the data sources. Provide additional descriptions as needed.

	Materials	Energy	Solid Waste	Emissions
Direct measurement				
Company purchasing/utility records				
Calculated from equipment specs				
Engineering estimates				
Permit limit				
Other:				

Has extrapolation or allocation been used during data collection? If so, in what way?

____

____

____

Figure 3.8 Life Cycle Inventory Data Collection Form (continued).

PROCESS NAME	Conventional Gas with EtOH Life Cycle Inventory -Summary
PROCESS ID	EtOH_wb.xls
REFERENCE FLOW	1014 gallons
PROCESS DESCRIPTION	Summary of the Gasoline with EtOH LCI using weight-based allocation.

BASIS OF CALCULATIONS

	Value	Units	Reference Website	Year	Basis
Number of Cars	1	Car	LCI Goal and Scope		
Distance Traveled per Year	20036	Miles	LCI Goal and Scope		

PROCESS INPUTS

	Value	Units	Reference Website	Year	Basis
Seed	6.35E+03	Kernels	Summary		
Nitrogen Fertilizer	3.36E+01	lb	Summary		
Potash Fertilizer	1.94E+01	lb	Summary		
Phosphate Fertilizer	1.40E+01	lb	Summary		
Lime	6.03E+01	lb	Summary		
Insecticides	1.14E-01	lb	Summary		
Herbicides	5.34E-01	lb	Summary		
Coal	1.63E+02	lb			
Crude Oil	8.87E+02	gal	Summary		
Natural Gas	5.13E+02	SCF			
Uranium	1.06E-01	lb	0		
Wood	6.51E+00	lb	0		
Water	8.97E+04	gal	0		
Drilling Fluids	Unknown		0		

PROCESS OUTPUTS

	Value	Units	Reference Website	Year	Basis
Air Emissions					
Water Emissions					
Solid Waste					
Fossil Fuel					
Non-Fossil Fuel					
Water Consumption					

Figure 3.9 Example Spreadsheet Reporting Life Cycle Inventory Data and Meta Data.

3.8 Life Cycle Inventory Data Quality

In practice, an LCI contains a mixture of measured, estimated, and calculated data. The quality of data is rarely homogenous with some being very reliable while others are estimated. In an attempt to improve how data quality is expressed, some attempt to mathematically aggregate data quality results to produce an overall multi-aspect score, first at the unit process level and then through the inventory and impact assessments. For example, ILCD data quality scores are used to develop an overall measure of data quality as an average score excluding those receiving a score of zero. This method implicitly values all data quality aspects equally. It is unclear that will not lead to misinterpretation. Is a score of 2 actually twice as good as a score of 4? Although desirable, especially for assessments intended to compare entire life cycles, such an "overall" data quality score does not provide a basis for concluding relative superiority among results across a large number of processes and flows (Cooper and Kahn 2012).

Instead of preparing an aggregate score, others consider data quality information as a contributor to the uncertainty of the data, e.g., using data quality scores to estimate the "additional" uncertainty from lower data quality. Weidema & Wesnaes (1996) developed a semi-quantitative, matrix approach using six types of Data Quality Indicators (DQIs) in a 1 - 5 scoring system (see Table 3.3) :

1. Reliability (related to the reliability of the collected primary data);
2. Completeness (related to the completeness of the primary data);
3. Temporal correlation (related to the temporal correlation of the primary data);
4. Geographical correlation (related to the geographical correlation of the secondary data used);
5. Further technological correlation (related to the technological correlation of the secondary data used)
6. Sample size (related to the sample size of the primary data).

Another example of data quality analysis comes from the US Department of Agriculture (USDA). In order to develop a system that allows easier and more consistent scoring among practitioners while still communicating what is good and bad about the data, a method has been proposed pursuant to the development of data for the USDA's LCA Digital Commons, an open access database and toolset (see http://www.lcacommons.gov/). The intent of the Digital Commons data quality method is to address the issue of data quality analysis reproducibility at the flow level. Using a two-tiered scale, presented in Table 3.4, each flow is assigned a score of A or B for each of the seven flow data quality indicators. A score of A indicates that the flow data meet the requirements; a score of B indicates that the unit process data do not meet the requirements. Preliminary use of the Digital Commons method appears to improve reproducibility, albeit at the expense of a finer data quality scale that could be useful during interpretation (Cooper and Kahn 2012).

Table 3.3 Semi-Quantitative Approach to Assessing Data Quality (Weidema and Wesnaes 1996).

DQI	1	2	3	4	5
Reliability	Verified data based on measurements	Verified data partly based on measurements OR non-verified data based on measurements	non-verified data partly based on qualified estimates	Qualified estimates (e.g. by industrial expert) data derived from theoretical information	Non-qualified estimate
Completeness	Representative data from all sites relevant for the market considered over an adequate period to even out normal fluctuations	Representative data from >50% of the sites market considered over an adequate period to even out normal fluctuations	Representative data from only some sites (<50%) relevant for the market considered OR >50% of the sites but from shorter periods	Representative data from only one site for the market considered OR some but from shorter periods	Representativeness unknown or data from a small number of sites AND from short periods
Temporal correlation	Less than 3 years of difference to reference year	Less than 6 years of difference to reference year	Less than 10 years of difference to reference year	Less than 15 years of difference to reference year	Age of data unknown OR more than 15 years difference from reference year
Geographical correlation	Data from area under study	Average data from smaller area than area under study or from similar area	Data from smaller area than area under study, or from similar area		Data from unknown or distinctly different area
Further technological correlation	Data from enterprises, processes and materials under study (i.e., identical technology)	Data from processes or materials under study (i.e. identical technology) but from different enterprises	Data from related processes or materials but same technology, OR data from processes and materials under study but from different technology OR process partially represented	Data from related processes or materials but different technology, OR data on laboratory scale processes and same technology	Data on related processes or materials but on laboratory scale of different technology
Sample size	>100, continuous measurement, balance of purchased, technical plans	>20	>10, aggregated figure in environmental report	≥3	unknown

Table 3.4 LCA Digital Commons Flow Data Quality Scoring Criteria (Cooper and Kahn 2012).

Data Quality Indicator	**Requirements for a Data Quality Score of A**
1. Reliability and Reproducibility	The flow data were based on measurements using a specified and standardized measurement method OR The flow data were estimated using methods and data described in specified archival or other consistently publically available sources.
2. Flow Data Completeness	The flow data were collected over at least 3 years for agricultural (crop, livestock, forest and range) processes or other processes in which the data point varies for uncontrolled annual conditions (e.g., weather) AND The flow data balance the mass and energy in and out of the unit process.[a]
3. Temporal Coverage	The flow data represent operations that occurred between the unit process start and end dates without forecasting.
4. Geographical Coverage	The flow data represent operations that occurred within the location of the unit process including nonagricultural process data that have been adapted to reflect logistics and market shares[b] for the unit process location.
5. Technological Coverage	The flow data represent the process(es) and/or material(s) specified without surrogacy or aggregation with other technologies.
6. Uncertainty	The flow data either include estimates of the first quartile, mean, median and third quartile values OR Data or probability distribution from which these values can be estimated.
7. Precision[c]	The relative standard error of the flow data is less than or equal to 25% OR The interquartile range divided by the median is less than or equal to 50% OR For a triangular distribution, the minimum flow data value is ≥75%, and maximum flow data value is ≤125% of the most likely value OR For a uniform distribution, the minimum flow data value is ≥75%, and maximum flow data value is ≤125% of the average of the minimum and maximum values.

[a] An incomplete mass balance may represent either an incomplete unit process or an incomplete set of emissions factors, or both. In the case of a score of B, e.g., for an incomplete set of emissions factors, the data quality analysis serves to highlight an opportunity to improve data quality through methodological or documentation improvement

[b] Market shares, sometimes called mixer processes in LCA, reflect the technologies used in local markets. For example, market shares are used to represent the mix of technologies used in regional electricity generation (the percentage of coal, natural gas, nuclear, etc. per kilowatt hour) and the mix of waste management technologies (landfilling, waste-to-energy, etc.) locally available

[c] In the precision category, percentages are intended to represent quartiles, as frequently used in descriptive statistics to represent a fourth of the population being sampled. Note also that for unit processes that balance in category 2, precision will apply as propagated to flows on both sides of the balance

3.9 Economic Input/Output (EIO) Data

Economic Input/Output (EIO) analysis is an economic discipline that models the interdependencies of production and consumption between industries and households within a nation's economy. The input/output model divides an entire economy into distinct sectors and represents them in table, or matrix, form so that each sector is represented by one row and one column (Table 3.5). The matrix represents sales from one sector to another. The economic input-output model is linear so the effects of purchasing $1,000 from one sector will be ten times greater than the effects of purchasing $100 from that sector.

The data models the economic flows (in millions of dollars) of goods throughout the economy and includes a matrix of close to 500 industrial sectors. It shows how the output of one industry is an input to other industries.

Today, almost all countries regularly compile IO tables to track their national accounts, although few are as detailed as the US model, which provides data across approximately 500 sectors. With the growth of eCommerce, price information for most commodities is available through an on-line search. The US Department of Commerce's Bureau of Economic Analysis provides IO tables for the United States (http://www.bea.gov/industry/).

Models which combine the economic input/output model with process models have been proposed to utilize the advantages offered by both approaches (Hendrickson, Lave *et al.* 2006). Merging EIO with LCA is sometimes referred to as EIOLCA, or hybrid LCA. It offers an alternative way to easily create LCI. To do so, the economic output for each sector is first calculated, and then the environmental outputs are calculated by multiplying the economic output at each stage by the environmental impact per dollar of output. The advantage of the economic input/output approach is that it quickly covers an entire economy, including all the material and energy inputs, thereby simplifying the inventory creation process. Its main disadvantage is the data are created at high aggregate levels for an entire industry, such as steel mills, rather than particular

Table 3.5 Simplified Economic Input/Output Data Available from the US Department of Commerce, Bureau of Economic Analysis (http://www.bea.gov/industry) $M.

Economic Activity	Inputs to Agriculture	Inputs to Manufacturing	Inputs to Transport	Final Demand	Total Output
Agriculture	5	15	2	68	90
Manufacturing	10	20	10	40	80
Transportation	10	15	5	0	30
Labor	25	30	5	0	60
Etc....	...	...	...	...	...

products, such as the type of steel used to make automobiles. Therefore, if the product being studies is representative of a sector, EIOLCA can provide a fast estimate of the complete supply chain implications.

EIOLCA methodology is a major research focus for the Green Design Institute at Carnegie Mellon University. Over the past 15 years, the group has investigated numerous products, services, and infrastructure systems using LCA as a fundamental component of analysis, leading field in EIO methodology and application, and produced an openly available on-line tool (http://www.eiolca.net/methods.html). Also, the Comprehensive Environmental Data Archive (CEDA) was created in the year 2000 by Sangwon Suh while at the Institute of Environmental Sciences (CML) at Leiden University. The newest version, CEDA 4.0, was released in 2009 using input-output tables and environmental statistics from 2002 (Suh 2009).

3.10 Consequential LCA

A consequential LCA is conceptually complex because it includes additional, economic concepts such as marginal production costs, elasticity of supply and demand, etc. Consequential LCA depends on descriptions of economic relationships embedded in models. It generally attempts to reflect economic relationships by extrapolating historical trends in prices, consumption and outputs. Some of the models are also much less transparent than the linear and static model of attributional LCA. Their results can also be very sensitive to the built-in assumptions. All these add to the risk that inadequate assumptions or other errors significantly affect the final LCA results. To reduce this risk, it is important to ensure that the various results regarding different consequences can be explained using credible arguments.

Consequential LCA is a clear example of how the deepening can be achieved in LCA. In fact, CLCA aims at describing the effect of changes within the life cycle and given that changes lead to a series of consequences through chains of cause-effect relationships. These mechanisms are at the core of this modeling technique (Curran *et al.* 2005). Present focus is on market mechanisms, i.e. those driven by the interaction of supply, demand, and prices. These mechanisms are dealt with exogenously[4] through the inclusion in the system under study of the affected processes, defined as those that respond to the change in demand driven by the decision at hand.

We consider the case of biofuels as a guiding example. The production of bioethanol represents a new energy product into the market, causing changes in prices and in volume within the energy market but also outside, affecting other commodities. Among the possible consequences, for example, it is possible that more corn would be required for bioethanol, squeezing out corn for food and also land use for wheat production. Both prices will increase, with still other products being squeezed out and rising in price. The chain of consequences to analyze could be very long and complex. However, in CLCA, simplifications are adopted, for example, in relation to the number of markets dealt with simultaneously, to the scale of consequences or to the complexity of substitution mechanisms.

---

[4] They are derived from economic models or outlooks in specific sector and then included as input into LCA.

Going beyond and thus not limiting the approach to the market mechanisms, the consequential logic could be extended so to consider CLCA not a modeling principle with defined rules but an approach to deepen LCA, to include more mechanisms. Which ones to include is a tricky question, since they can show up everywhere, involving a variety of domains. Market mechanisms are part of broader economic mechanisms, which are related to concepts like employment and growth. These in turn function within a cultural, social, political and regulatory context. Taking this complex chain of consequences into account would require important developments in CLCA, the main one being the removal of those constraints that presently are considered as fixed entities. Thus, in perspective CLCA offers already the conceptual basis for a life cycle sustainability assessment, proposing itself as a way for dealing with the modeling step in the LCSA framework. However, it is necessary first to improve present capabilities of CLCA to fully address market mechanisms, before extending it to a more complex analysis. Efforts are necessary to clarify when and which market information is important, to improve long-term forecasting techniques and to identify the affected processes. In this regard, a great contribution could be given by scenario modeling, since it would provide a more sound-scientific basis to model specific product-related futures for example with respect to technology development and market shifts.

It is possible that the inventory results of a consequential LCA will be negative, if the change in the level of production causes a reduction in emissions greater than the emissions from the production of the product. This does not mean that the absolute emissions from the production of the product are negative, but that the production of the product will cause a reduction in emissions elsewhere in the system.

In the end, both approaches are legitimate and fulfill different needs (Ekvall *et al.* 2005). The distinction between attributional and consequential LCA is one example of how choices in the Goal and Scope Definition of an LCA should influence methodological and data choices for the LCI and LCIA phases.

3.11 LCA Software

Conducting an LCA relies heavily on both data and software. A typical LCA case study requires information about hundreds or thousands of processes and their input and output flows, and all these processes need to be connected, and a balance be calculated. When generating the product system and life cycle model, users appreciate, and often need, a convenient display of connected processes, methods for error checks, quick calculation of intermediate results, data import and export abilities, and so forth.

Consequently, for more than 20 years, a market for professional LCA software[5] has been established. Earlier approaches often relied on spreadsheet software that does not meet many of the criteria mentioned above. Currently, some working groups use mathematical modeling software such as Matlab[6].

LCA software has a high influence on the approaches that are possible and easily available for performing an LCA; software "transforms" data into a model, and provides

---

[5] LCA software is simply understood, in this chapter, as software that is related to Life Cycle Assessment

[6] For example, the industrial ecology group at NTNU in Trondheim, Norway, www.ntnu.no/indecol

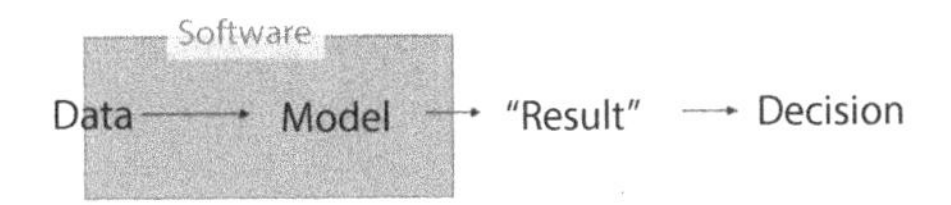

Figure 3.10 LCA software "transforming" data via a model that is generated within the software to a result that is then finally used to support a decision (Ciroth 2006).

results to the user that are then used, generally, in decision support. In other words, there are no LCA results without software.

There are several commercially-available LCA software packages. In recent years, GaBi from PE International (www.gabi-software.com), and SimaPro from PRé Consultants (www.pre.nl/simapro) probably[7] have the highest market share on a worldwide level. Both products have been on the market for more than fifteen years. While each has a different "style" in modeling a product system offers different features, both have a broad number of features in common. Umberto, by ifü Hamburg, also offers LCA modeling that is somewhat more complex than with SimaPro or GaBi. An overview of available LCA modeling software is provided by the European Commission's Joint Research Center, which is occasionally updated (see Table 3.6).

3.11.1 Characteristics of LCA Software Systems

LCA software systems differ a lot – there are complete systems and LCA software that is an add-on to other software, there are closed and open source tools, and so forth. The following sections explore some of the main differences in detail.

3.11.2 Web Tools versus Desktop Tools

Traditionally, software had to be installed locally, on a desktop; using it required access to this desktop computer. Since the emergence of the internet, software can also be installed on a web server, and accessed by any user that has access to this web server. The web server does not need to be publicly available on the internet, a local, in-house server is possible too.

Recently, several LCA packages have been launched as web application, for example greenfly (www.greenflyonline.org/, accessed January 30, 2015) in Australia or the Quantis Suite by Quantis international (Quantis 2012). With e-DEA, a web application add-on to classic, desktop or client-server LCA software systems has been created. e-DEA links to SimaPro, SimaPro can be installed on a local server or on a web-server, or even on a desktop (www.edea-software.com/, accessed January 30, 2015).

Online-tools usually offer a clean, modern, and therefore attractive user interface[8]. Furthermore, they need to be installed only once, which is an advantage for larger companies or larger user groups; finally, data can be centrally managed, and software updates can be managed centrally.

---

[7] There are no official statistics, nor do the companies publish data on their sales.

[8] One reason may simply be that they are relatively new and, therefore, better reflect recent user interface design conventions and experience.

Table 3.6 European Commission's Joint Research Centre (JRC) List of Tools. http://eplca.jrc.ec.europa.eu/ResourceDirectory/toolList.vm (accessed 29 January 2015).

Tool + version N°	Developer	Supplier	Instruments	ILCD Compliant	ILCD Entry Level	PEF/OEF	Languages of Interface
EIME	Bureau Veritas CODDE	Bureau Veritas CODDE	Life cycle impact assessment (LCIA), Life cycle inventory (LCI), Life cycle assessment (LCA), Legal Complience checks, Life cycle costing (LCC)	Yes	No	No	English
GaBi DfX	PE INTERNATIONAL	FEBE ECOLOGIC PE INTERNATIONAL	social LCA, Life cycle inventory (LCI), Life cycle impact assessment (LCIA), Life cycleengineering (LCE), Substance/material flow analysis (SFA/MFA), Disassembly modeling, Life cycle management (LCM), Life cycle assessment (LCA), Design for environment (DfE, DfR), Product stewardship, supply chain management, Legal Complience checks, Life cycle sustainability assessment (LCS), Life cycle costing (LCC)	Yes	No	No	Japanese, Spanish Portuguese, French, Thai, Italian, Chinese, German, English
SimaPro	ESU-services Ltd	ESU-services Ltd		Yes	Yes	No	French, German, English

Umberto NXT LCA	ifu Hamburg GmbH	ifu Hamburg GmbH	Life cycle management (LCM), Life cycle impact assessment (LCIA), Life cycle inventory (LCI), Life cycleengineering (LCE), Life cycle assessment (LCA), Substance/material flow analysis (SFA/MFA). Product stewardship, supply chain management, Life cycle sustainability assessment (LCS), Life cycle costing (LCC)	Yes	No	No	English
AIST-LCA Ver.4	National Institute of Advanced Industrial Science and Technology (AIST)	National Institute of Advanced Industrial Science and Technology (AIST)	Life cycle management (LCM), Life cycle impact assessment (LCIA), Life cycle Inventory (LCI), Life cycle assessment (LCA), Product stewardship, supply chain management.	No	No	No	Japanese
BEE (Bilan Environnemental des Emballages)	Eco-Ernballages		Life cycle impact assessment (LCIA), Life cycle inventory (LCI), Life cycle assessment (LCA), Design for environment (DfE, DfR)	No	No	No	French, English
BEES 3.0d	National Institute of Standards and Technology (NIST)	National Institute of Standards and Technology (NIST)	Life cycle impact assessment (LCIA), Life cycle inventory (LCI), Life cycle assessment (LCA), Life cycle costing (LCC)	No	No	No	English

(*Continued*)

Table 3.6 (Cont.)

Tool + version N°	Developer	Supplier	Instruments	ILCD Compliant	ILCD Entry Level	PEF/OEF	Languages of Interface
DPL 1.0	IVAM University of Amsterdam bv	IVAM University of Amsterdam bv		No	No	No	Dutch
e!Sankev	ifu Hamburg GmbH	ifu Hamburg GmbH	Life cycle inventory (LCI) Life cycleengineering (LCE), Life cycle assessment (LCA), Substance/material flow analysis (SFA/MFA), Design for environment (DfE, DfR), Life cycle sustainability assessment (LCS). Life cycle casting (LCC)	No	No	No	Spanish, Portuguese, French, German, English
e-LICCO	CYCLECO	CYCLECO	Life cycle assessment (LCA)	No	No	No	French
Eco-Bat 2.1	Haute Ecole d'lngénierie et de Gestion du Canton de Vaud	Haute Ecole d'lngénierie et de Gestion du Canton de Vaud	Life cycle impact assessment (LCIA), Design for environment (DfE, DfR)	No	No	No	French, Italian, English
Eco-Quantum	IVAM University of Amsterdam bv	IVAM University of Amsterdam bv		No	No	No	Dutch
ECODESIGN X-Pro v1.0	EcoMundo	EcoMundo	Life cycle impact assessment (LCIA), Life cycle inventory (LCI), Life cycle assessment (LCA), Legal Complience checks	No	No	No	English

ecoinvent waste disposal inventory tools v1.0	Doka Life Cvcle Assessments (Doka Okobilanzen)	Doka Life Cvcle Assessments (Doka Okobilanzen)	Life cycle inventory (LCI)	No	No	No	English
EcoScan 3.1	TNO Built Environment & Geosciences	TNO Built Environment & Geosciences	Life cycle impact assessment (LCIA), Design for environment (DfE, DfR)	No	No	No	Spanish, German, Dutch, English
Environmental Impact Estimator V3.0.2	Athena Sustainable Materials Institute	Athena Sustainable Materials Institute	Life cycle impact assessment (LCIA), Life cycle assessment (LCA). Design for environment (DfE, DfR)	No	No	No	English
EPD Tools Suit 2007	IKE Environmental Technology	IKE Environmental Technology	Life cycle inventory (LCI)	No	No	No	Chinese
eVerdEE v.1.0	ENEA - Italian National Agency for New Technology. Energy and the Environment	ENEA - Italian National Agency for New Technology. Energy and the Environment	Life cycle management (LCM), Life cycle assessment (LCA), Design for environment (DfE, DfR)	No	No	No	Spanish, Italian, German, English
eVerdEE v.2.0	ENEA - Italian National Agency for New Technology. Energy and the Environment	ENEA - Italian National Agency for New Technology. Energy and the Environment	Life cycle management (LCM), Life cycle assessment (LCA), Design for environment (DfE, DfR)	No	No	No	Italian, English
Food'Print	CYCLECO	CYCLECO	Life cycle assessment (LCA)	No	No	No	French, English

(*Continued*)

Table 3.6 (Cont.)

Tool + version N°	Developer	Supplier	Instruments	ILCD Compliant	ILCD Entry Level	PEF/OEF	Languages of Interface
GaBi Envision	PE INTERNATIONAL	FEBE ECOLOGIC PE INTERNATIONAL	Life cycle management (LCM), Life cycle impact assessment (LCIA), Life cycle Inventory (LCI), Life cycle assessment (LCA), Substance/material flow analysis (SFA/MFA). Design for environment (DfE, DfR), Product stewardship, supply chain management	No	No	No	Spanish, Portuguese, English
GaBi Software	PE INTERNATIONAL LBP. University of Stuttgart (former IKP)	FEBE ECOLOGIC PE INTERNATIONAL LCA Center Denmark	social LCA, Life cycle management (LCM), Life cycle impact assessment (LCIA), Life cycle inventory (LCI), Life cycleengineering (LCE), Life cycle assessment (LCA), Substance/material flow analysis (SFA/MFA), Design for environment (DfE, DfR), Legal Complience checks, Product stewardship, supply chain management. Life cycle sustainability assessment (LCS), Life cycle costing (LCC), Other - please specify	No	No	No	Japanese, Spanish, Portuguese, French, Thai, Italian, Chinese, German, English

GEMIS version 4.4	Oeko-lnstitut (Institute for applied Ecology) Darmstadt Office	Oeko-lnstltut (Institute for applied Ecology Darmstadt Office		No	No	No	Spanish, Czech, German, English
Green-E. verslon 10	Quantis - Sustainability counts	Quantis - Sustainability counts	social LCA, Life cycle management (LCM), Life cycle assessment (LCA), Design for environment (DfE, DfR), Life cycle sustainability assessment (LCS) Life cycle costing (LCC)	No	No	No	English
KCL-ECO 4.0	Oy Keskuslaboratorio-Centrallaboratorium Ab KCL	Oy Keskuslaboratorio-Centrallaboratorium Ab KCL	Life cycle management (LCM), Life cycle impact assessment (LCIA), Life cycle Inventory (LCI), Life cycleengineering (LCE), Life cycle assessment (LCA), Substance/material flow analysis (SFA/MFA), Design for environment (DfE, DfR), Product stewardship, supply chain management	No	No	No	English
Key parameter model for energy systems	ESU-services Ltd.	ESU-services Ltd.		No	No	No	French, German

(*Continued*)

Table 3.6 (Cont.)

Tool + version N°	**Developer**	**Supplier**	**Instruments**	**ILCD Compliant**	**ILCD Entry Level**	**PEF/OEF**	**Languages of Interface**
LCA – Evaluator 2.0	GreenDeltaTC GmbH	GreenDeltaTC GmbH	Life cycle management (LCM), Life cycle impact assessment (LCIA), Life cycle assessment (LCA)	No	No	No	English
LEGEP1.2	LEGEP Software GmbH	LEGEP Software GmbH	social LCA, Life cycle management (LCM), Life cycle impact assessment (LCIA), Life cycleengineering (LCE), Life cycle assessment (LCA), Design for environment (DfE, DfR), Life cycle sustainability assessment (LCS), Life cycle costing (LCC)	No	No	No	Italian, German
LTE OGIP; Version 5.0; Build-Number 2092: 2005/12/12	t.h.e. Software GmbH	t.h.e. Software GmbH	Life cycle impact assessment (LCIA), Life cycle inventory (LCI), Life cycle assessment (LCA), Design for environment (DfE, DfR), Life cycle costing (LCC)	No	No	No	German
Mi LCA	Japan Environmental Management Association for Industry (JEMAI)	Japan Environmental Management Association for Industry (JEMAI)	Life cycle impact assessment (LCIA), Life cycle inventory (LCI), Life cycle assessment (LCA)	No	No	No	Japanese, English

Modular MSWI Model 1.0	GreenDeltaTC GmbH	GreenDeltaTC GmbH	Life cycle management (LCM), Life cycle impact assessment (LCIA), Life cycle Inventory (LCI), Life cycleengineering (LCE), Life cycle assessment (LCA), Substance/material flow analysis (SFA/MFA), Design for environment (DfE, DfR), Life cycle costing (LCC)	No	No	No	English
Prototype Demolition Waste Decision Tool 1	IVAM University of Amsterdam bv	IVAM University of Amsterdam bv	Life cycle impact assessment (LCIA), Life cycle inventory (LCI), Life cycle assessment (LCA), Design for environment (DfE, DfR), Product stewardship, supply chain management	No	No	No	Dutch
REGIS 2.3	sinum AG	sinum AG	Life cycle management (LCM), Life cycle impact assessment (LCIA), Life cycle inventory (LCI), Life cycle assessment (LCA), Substance/material flow analysis (SFA/MFA), Legal Complience checks, Life cycle sustainability assessment (LCS), Life cycle costing (LCC)	No	No	No	Japanese, Spanish, German, English

(*Continued*)

Table 3.6 (Cont.)

Tool + version N°	Developer	Supplier	Instruments	ILCD Compliant	ILCD Entry Level	PEF/OEF	Languages of Interface
SALCA-animal 1.0	Agroscope Reckenholz-Tänikon Research Station ART	Agroscope Reckenholz-Tänikon Research Station ART	Life cycle inventory (LCI)	No	No	No	German
SALCA- biodiversity 061	Agroscope Reckenholz-Tänikon Research Station ART	Agroscope Reckenholz-Tänikon Research Station ART	Life cycle impact assessment (LCIA), Life cycle inventory (LCI)	No	No	No	German
SALCA-biodiversity 1.0	Agroscope Reckenholz-Tänikon Research Station ART	Agroscope Reckenholz-Tänikon Research Station ART	Life cycle impact assessment (LCIA), Life cycle inventory (LCI)	No	No	No	German
SALCA-crop 061	Agroscope Reckenholz-Tänikon Research Station ART	Agroscope Reckenholz-Tänikon Research Station ART	Life cycle impact assessment (LCIA), Life cycle inventory (LCI), Life cycle assessment (LCA)	No	No	No	German
SALCA-crop 2.02	Agroscope Reckenholz-Tänikon Research Station ART	Agroscope Reckenholz-Tänikon Research Station ART	Life cycle impact assessment (LCIA), Life cycle inventory (LCI), Life cycle assessment (LCA)	No	No	No	German
SALCA-erosion 061	Agroscope Reckenholz-Tänikon Research Station ART	Agroscope Reckenholz-Tänikon Research Station ART	Life cycle inventory (LCI)	No	No	No	German

SALCA-erosion 2.0	Agroscope Reckenholz-Tänikon Research Station ART	Agroscope Reckenholz-Tänikon Research	Life cycle inventory (LCI)	No	No	No	German
SALCA-farm 1.31	Agroscope Reckenholz-Tänikon Research Station ART	Agroscope Reckenholz-Tänikon Research Station ART	Life cycle impact assessment (LCIA), Life cycle inventory (LCI), Life cycle assessment (LCA)	No	No	No	German
SALCA-farm 2.1	Agroscope Reckenholz-Tänikon Research Station ART	Agroscope Reckenholz-Tänikon Research Station ART	Life cycle impact assessment (LCIA), Life cycle inventory (LCI), Life cycle assessment (LCA)	No	No	No	German
SALCA-heavy metals 061	Agroscope Reckenholz-Tänikon Research Station ART	Agroscope Reckenholz-Tänikon Research Station ART	Life cycle inventory (LCI)	No	No	No	German
SALCA-heavy metals 1.0	Agroscope Reckenholz-Tänikon Research Station ART	Aqroscope Reckenholz-Tänikon Research Station ART	Life cycle inventory (LCI)	No	No	No	German
SALCA-nitrate 061	Agroscope Reckenholz-Tänikon Research Station ART	Agroscope Reckenholz-Tänikon Research Station ART	Life cycle inventory (LCI)	No	No	No	German
SALCA-nitrate 4.0	Agroscope Reckenholz-Tänikon Research Station ART	Agroscope Reckenholz-Tänikon Research Station ART	Life cycle inventory (LCI)	No	No	No	German
SALCA-soil quality 061	Agroscope Reckenholz-Tänikon Research Station ART	Agroscope Reckenholz-Tänikon Research Station ART	Life cycle impact assessment (LCIA), Life cycle inventory (LCI)	No	No	No	German

(*Continued*)

Table 3.6 (Cont.)

Tool + version N°	Developer	Supplier	Instruments	ILCD Compliant	ILCD Entry Level	PEF/OEF	Languages of Interface
SALCA-soil quality 1.1	Aqroscope Reckenholz-Tänikon Research Station ART	Agroscope Reckenholz-Tänikon Research Station ART	Life cycle impact assessment (LCIA), Life cycle inventory (LCI)	No	No	No	German
SankeyEdltor 3.0	STENUM GmbH	STENUM GmbH		No	No	No	English
SimaPro 7	PRé Consultants B.V.	2B PRé Consultants B.V.	social LCA, Life cycle management (LCM), Life cycle impact assessment (LCIA), Life cycle inventory (LCI), Life cycleengineering (LCE), Life cycle assessment (LCA), Substance/material flow analysis (SFA/MFA), Design for environment (DfE, DfR), Product stewardship, supply chain management, Life cycle sustainability assessment (LCS), Life cycle costing (LCC)	No	No	No	Japanese, Spanish, Danish. Greek, French. Italian, German, Dutch, English
Spin'it	CYCLECO	CYCLECO	Life cycle inventory (LCI). Life cycle assessment (LCA)	No	No	No	French, German, English

STAN 1.1.3-Software for Substance Flow Analysis	Vienna University of Technology	Vienna University of Technology	Substance/material flow analysis (SFA/MFA)	No	No	No	German, English
TEAM™ (Tool for Environmental Analysis and Management	Ecobilan - Pricewater-houseCoopers	Ecobilan - Pricewater-houseCoopers	Life cycle management (LCM), Life cycle impact assessment (LCIA), Life cycle Inventory (LCI), Life cycle assessment (LCA), Substance/material flow analysis (SFA/MFA), Design for environment (DfE. DfR), Legal Complience checks, Product stewardship, supply chain management, Life cycle costing (LCC)	No	No	No	English
TESPI	ENEA - Italian National Agency for New Technology. Energy and the Environment	ENEA - Italian National Agency for New Technology. Energy and the Environment	Design for environment (DfE, DfR)	No	No	No	Italian, English
The Boustead Model 5.0.12	Boustead Consulting Limited	Boustead Consulting Limited	Life cycle impact assessment (LCIA), Life cycle inventory (LCI), Life cycle assessment (LCA)	No	No	No	English

(*Continued*)

Table 3.6 (Cont.)

Tool + version N°	Developer	Supplier	Instruments	ILCD Compliant	ILCD Entry Level	PEF/OEF	Languages of Interface
trainEE	Green DeltaTC GmbH	GreenDeltaTC GmbH	Life cycle management (LCM), Life cycle impact assessment (LCIA), Life cycle inventory (LCI), Life cycleengineering (LCE), Life cycle assessment (LCA), Substance/material flow analysis (SFA/MFA), Design for environment (DfE, DfR), Product stewardship, supply chain management, Life cycle costing (LCC)	No	No	No	English
Umberto NXT CO2	ifu Hamburg GmbH	ifu Hamburg GmbH	Life cycle management (LCM), Life cycle impact assessment (LCIA), Life cycle inventory (LCI), Life cycleengineering (LCE), Life cycle assessment (LCA), Substance/material flow analysis (SFA/MFA), Product stewardship, supply chain management, Life cycle sustainability assessment (LCS), Life cycle costing (LCC)	No	No	No	German, English

USES-LCA	Radboud University Nijmegen	Radboud University Nijmegen	Life cycle impact assessment (LCIA)	No	No	No	English
Verdee	ENEA - Italian National Agency for New Technology. Energy and the Environment	ENEA - Italian National Agency for New Technology. Energy and the Environment	Life cycle management (LCM), Design for environment (DfE, DfR)	No	No	No	Italian
WAMPS. betaversion	IVL 1Swedish Environmental Research Institute Ltd	IVL Swedish Environmental Research Institute Ltd		No	No	No	English
WRATE	UK Environment Agency	UK Environment Agency	Life cycle management (LCM), Life cycle impact assessment (LCIA), Life cycle inventory (LCI), Life cycle assessment (LCA), Life cycle sustainability assessment (LCS)	No	No	No	English

At first, web applications lacked many of the features of desktop LCA software system. But now there seems to be a trend to enrich web applications. For example, the Quantis Suite 2.0 allows the user to select from several impact assessment methods, while previously only one impact assessment method was available.

The disadvantage of web tools is that the local user has less control over software and data, and typically also over the server where the software is running; if the server is not responding, either because the server is down or because the connection is not working, the software cannot be used. Users might be reluctant to upload sensitive data to a web server that is not fully under their control. Also, handling large amounts of data requires careful optimization even with modern, fast connections, and is still posing challenges to web applications.

Multi-user desktop software systems are in some respects comparable to web-based software; also here, data can be centrally managed. Software needs to be installed locally, though, too. SimaPro for example is available also in a multi-user, client/server license type.

3.11.3 Commercial Tools versus Freeware

With the emergence of LCA methodology, also license fees for LCA software systems have increased, to several thousand Euros per license. Yet still, some tools have always been freely available for example:

- CMLCA, created by Reinout Heijungs at CML Leiden (www.cmlca.eu, accessed January 30, 2015), and
- Gemis, hosted by International Institute for Sustainability Analysis and Strategy (IINAS) (http://www.iinas.org/gemis-download-en.html, accessed January 30, 2015).

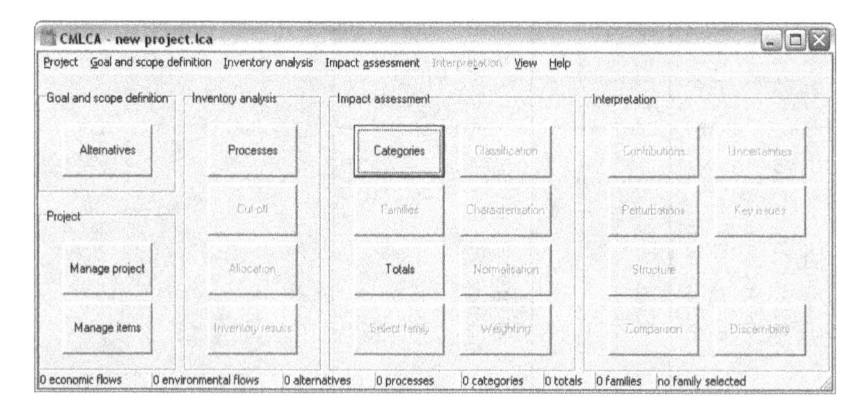

Figure 3.11 Screenshot of the CMLCA User Interface.

These free tools usually do not offer all features of commercial LCA systems; workflow and user interface design do not always comply with user interface conventions, and sometimes are different on purpose. More advanced features, as the graphical

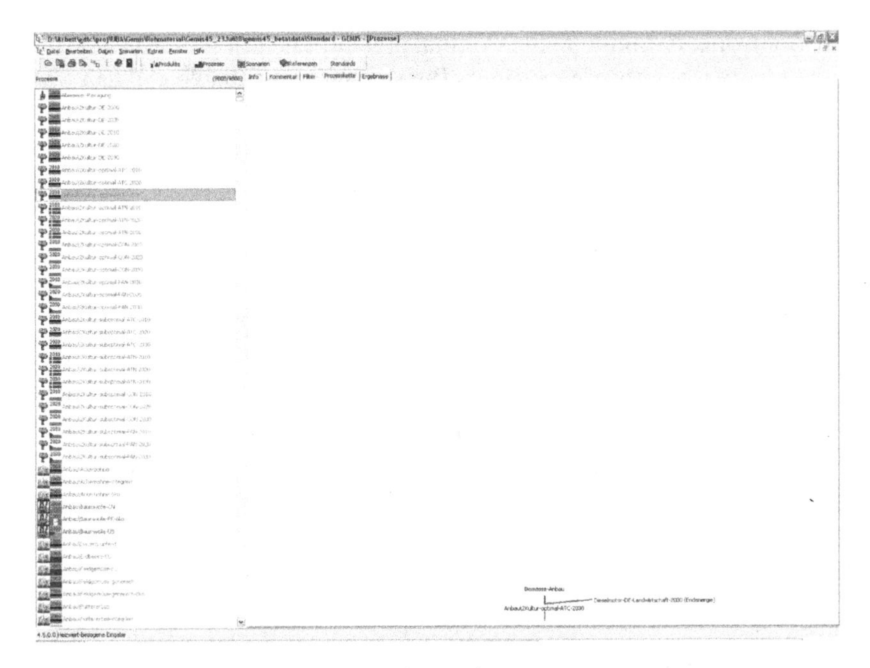

Figure 3.12 Screenshot Gemis – Process Network Visualisation.

visualisation of supply chains, and in-depth result and contribution analysis may be lacking. However, some more advanced and special, rather experimental, methods may be available in free tools.

Open source tools are also freely available; they are different from the free tools discussed here since they can also have a commercial side.

3.11.4 Open Source versus Closed Source

"Open source" is a specific term developed for software; open source software is released under a specific open source license. There are many different open source licenses, most are listed at www.opensource.org. Main characteristics of open source licenses are:

1. The source code of the software is available to everyone, for free, and
2. The license that accompanies every distributed file of software recognizes the originator of the code, and provides a mechanism to further ensure the open source "nature" of the file.[9]

openLCA, by GreenDeltaTC, is an internationally used, open source software program (openLCA 2012). The first version was released in September 2008; the sixth release is available at http://openlca.org.

Open source approaches provide several advantages to software users: First, the software is free to use; second, it is fully transparent, and third, it can be updated and

[9] There are more specific requirements. The website "www.opensource.org" contains a full list and explanations. Licenses that do not fulfill these requirements cannot be called open source licenses.

modified by anyone, not only by the original creator. The last point makes users more independent from the software provider. The software is available and can be updated even if the original creator has lost interest. For the creator of the software, a broad range of other open source software packages is available that can be combined or adapted to fit to the own software; this allows building a more mature and feature-rich software comparatively fast.

Although open source software is provided for free to users, it can be a viable business model for the software creator (Ciroth 2008): The benefit of available other open source software that can be integrated and reused, an existing infrastructure for distributing open source software, and a more quickly growing user base (with requests for additional features, tailored versions, or simply training) can outweigh the license fees.

3.11.5 General LCA Tools versus Specialized Tools versus Add-Ons

The purpose of LCA software may be quite different; on one hand, there are full LCA modeling systems that offer users all features required for performing complete LCA case studies. These are the main focus of this paper, and will be investigated later more in detail.

On the other hand, there are specialized tools, for example:

- The ecoeditor (http://www.ecoinvent.org/ecoinvent-v3/ecoeditor-v2/) provided by the ecoinvent centre, edits ecoSpold data sets and for managing the review process of these data sets.
- The ILCD editor (lca.jrc.ec.europa.eu/ieditor/ilcd_editor.jnlp) edits data sets in ILCD format.
- The openLCA data format converter (www.greendeltatc.com/openLCA-Format-Converter.117.0.html?&L=1) converts data sets from one LCA format into the other

Then, there are tools targeting only a specific user group; these tools are often created in dedicated research projects, providing for example a specific technical language for the targeted users, specific databases, and/or specific modeling choices.

Siegenthaler *et al.* (2005, p. 8) list specific LCA tools for the electric/electronic sector, for the building, waste, and pulp and paper industry. As the online list at JRC shows, there is an ongoing trend for creating specific tools.

Finally, closely related to the last group, are software add-ons that provide some LCA calculation or modeling features as add-on to software that is used mainly for other purposes.

One of the first of these examples is the SolidWorks "Sustainability Xpress" extension that was also extensively announced in videos, conferences, and other media[10]. Solidworks is a professional CAD software (SolidWorks 2012). The SolidWorks extension adds values for energy, climate change, "air" (pollution), and "water" (pollution). These values are calculated in advance. From a scientific standpoint, "carbon" and "energy" might be seen as closely related, and some other aspects relevant to the sustainability of a product are omitted (toxic aspects, land use, social effects). Also, the

[10] E.g. http://www.youtube.com/watch?v=Ts3EtsfM0QM

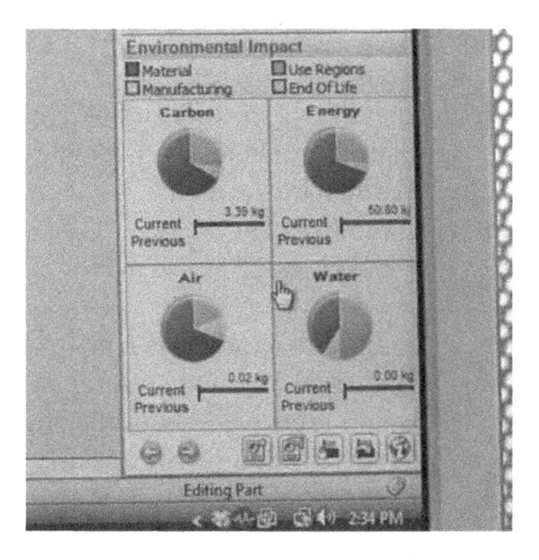

Figure 3.13 SolidWorks Sustainability XPress, Screenshot from youtube Video.

total impacts of a product can hardly be determined in the design phase (where a CAD software is used) since often the specific use of the product is responsible for a large share of the impact. But still, relevant aspects of LCA are put in front of the product designer, who then sees directly the impact of product design decisions on the environmental impact, and can directly consider it.

Meanwhile, a more comprehensive extension is released; also, other solutions that target non-LCA users are available, for example e-DEA (e-DEA 2012). E-DEA calculates results on the fly and allows real, more comprehensive modeling, but is not part of another software product; rather, it is linked to SimaPro, using it as data repository and calculation engine.

Examples for fully-featured LCA software systems are GaBi, SimaPro, Umberto, and openLCA.

3.11.6 Two Basic LCA Software User Types and Their Needs

LCA software *user types* are usually not discussed in the different LCA software assessment reports, e.g., Menke *et al.* 1996; Dunmade 2007; Siegenthaler *et al.* 2005, Jönbrink *et al.* 2000. The many different and partially conflicting characteristics of LCA software indicate that this would probably deserve an own investigation. For this chapter, however, only two different user types are distinguished[11]:

- Professional LCA modelers: This user type actively creates new LCA models, using all required methodological aspects of LCA;
- LCA model users: This user type uses existing LCA models in software.

[11] (Unger *et al.* 2004) define two similar LCA software user types, "scientists and research" (similar to the LCA modelers above) and "industry" (similar to LCA model users); however, since LCA has disseminated in many different areas, there are, for example, now also industry users modeling comprehensive, detailed LCA models, for business reason and not primarily for research.

LCA model users require[12]:

1. Quick, intuitive understanding of the tool and of data in the tool;
2. Easy and fast use, especially it should be easy and quick to come to results;
3. Reliable information provided by the tool;
4. More detailed information and help available as backup (for some few cases where model and tool are not as expected for example)

These points seem better met by web applications, or by add-on applications that link to software systems that users already, or anyhow, use.

Professional LCA modelers would probably not decline one of these aspects – however, the following points are relevant in addition for this user type:

1. Ability to model LCA systems with all methodological details (parameterization; allocation, system expansion);
2. Graphical display of the modeled product system;
3. Drill-down result analysis, ability to identify hot spots and the contribution of single elements to the overall result:
4. Speed, swiftness, usability
5. User community, support for users
6. Relevant LCA data should be available for the software
7. Interoperability, ability to import and export

Professional modelers will accept a learning curve for the software; currently, most LCA software houses offer a training of several days to train users in using the software efficiently.

Between these two basic user types, there is a transition zone – for example, LCA modelers might work with aggregated system processes for background processes that do not allow a detailed analysis of single contributions, and focus on the foreground.

3.11.7 The LCA Software Market

A market for LCA software developed quite early with the "emergence" of LCA, probably due to the high data demands and the rather specific requirements of the method. The market is quite dynamic, with new products entering and vanishing every year. For example, five of the ten LCA Software packages compared by Dunmade (2007) are not available today, including Cumpan by (formerly) Daimler Benz Inter Services.

For several years, two software systems have dominated the market: SimaPro and GaBi. Each offers the functionality described above as desirable (more or less, with different interpretation of this requirement). They will be described in more detail in the following sections. There are plenty other software systems, some of which will be highlighted as well.

---

[12] Source is mostly the experience of Andreas Ciroth, GreenDelta.

3.11.8 The Main LCA Software Systems

In recent years, GaBi from PE International (www.gabi-software.com), and SimaPro from PRé Consultants (www.pre.nl/simapro) are the two commercial LCA software packages with probably the highest market share on a worldwide level[13]. Both GaBi and SimaPro have been on the market for more than fifteen years. While both have a different "style" in modeling a product system, and have features that the competitor does not offer, they both have a broad number of features in common, albeit not all features are available in the basic (cheapest) license versions.

Prices for both SimaPro and GaBi are in the range of several thousand Euros, with academic licenses offered at a considerable lower price. An international network of resellers offers local support for each of the systems.

Picking only one point as a main differentiation, SimaPro calculates a product system in a matrix inversion, using thereby a highly efficient algorithm that allows dealing with thousands of processes in one calculation. GaBi, on the other hand, uses a sequential calculation algorithm that literally goes from one process to the other to scale the process according to its input to the overall LCA system.

In consequence, SimaPro is able to deal with a huge number of unit processes in one calculation; the contribution of each modeling step to the end result is available only after the calculation. The algorithm allows SimaPro to use only unit processes in a calculation, following up the links of one process to another dynamically, during calculation (Figure 3.14). Due to the dynamic linking, results always represent the recent database status. Updates of processes in the database are automatically reflected in the results. Loops in processes (steel as input into the steel making process for example) do not pose problems in the calculation.

In GaBi, users manually build a system of connected processes. This usually takes more time, but since the result is calculated stepwise, users get feedback on single modeling steps, creating modeled pieces, putting them together like building blocks (Figure 3.15). Processes can have more than one product, since they do not need to fit into a matrix. Calculating thousands of processes in one run can pose problems, with loops in the data, too. Updating the processes when reusing a once modeled process chain is the responsibility of the user; selected processes are constantly linked to the model, and updates in the data base are not automatically reflected. On the other hand, model results remain stable even if background data are updated.

The "case" with SimaPro and GaBi is an illustration of the importance of software for the application of LCA. GaBi process models tend to rely on aggregated system processes, while SimaPro relies on unit processes; unit processes in SimaPro must have only one product, which must be an output of the process; even waste treatment processes therefore are modeled to have an output product "waste treatment service" that is input to the process creating waste.

[13] There are no official statistics, nor do the companies publish data on their sales.

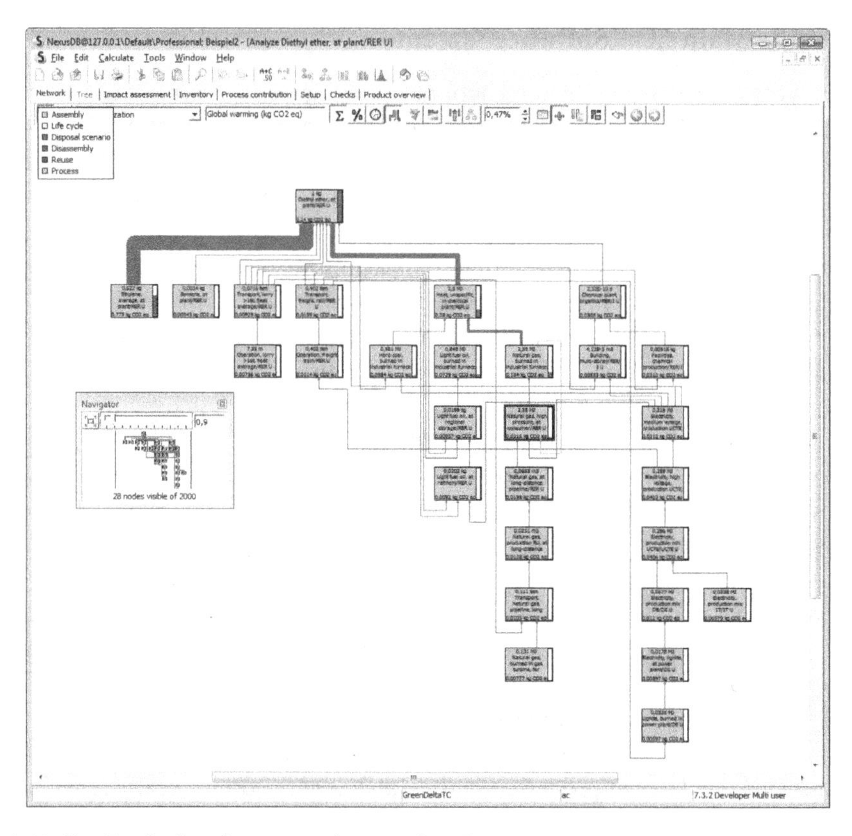

Figure 3.14 SimaPro Sankey diagram results view, SimaPro 7.3.2.

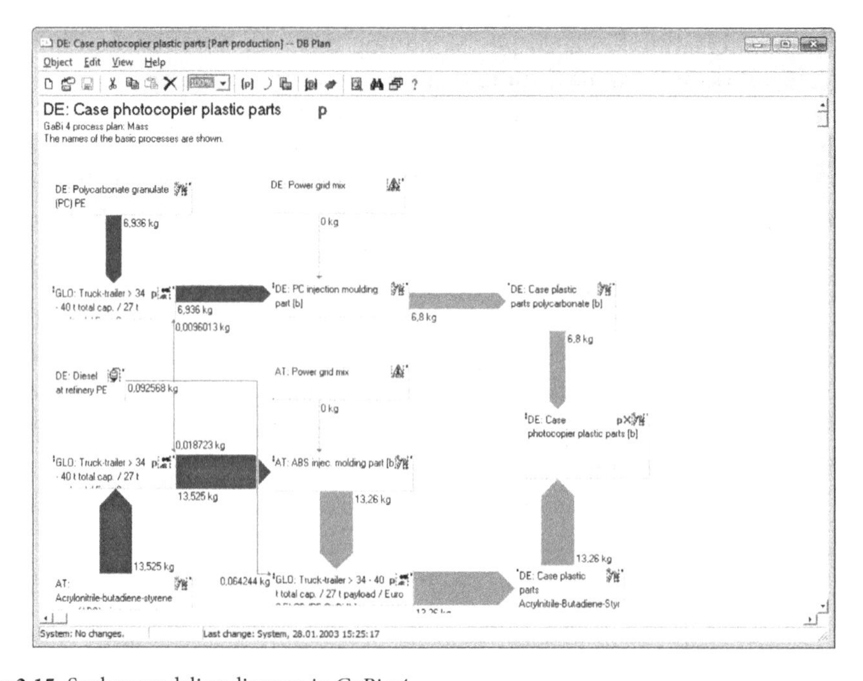

Figure 3.15 Sankey modeling diagram in GaBi v4.

References

Boguski TK (2000) Evaluation of Public Databases as Sources of Data for Life Cycle Assessments. US Environmental Protection Agency, Office of Research and Development, Cincinnati, Ohio, USA.

Ciroth A (2008) Business models for open source projects in environmental informatics, presentation and conference proceedings, Envirolnfo 2008, 12 September 2008

Ciroth A (2006) A New Open Source LCA Software, in Presentation and Proceedings, EcoBalance Conference Tsukuba, 14-16 November 2006.

Consoli F, Allen D *et al.* eds (1993) Guidelines for Life-Cycle Assessment: A Code of Practice, The Society of Toxicology and Chemistry (SETAC).

Curran MA (2011) Maintaining Quality Critical Peer Review (CPR) as the Demand for Life Cycle Assessments Increases. Life Cycle Management (LCM 2011) Conference, Berlin, Germany.

Curran MA and Notten P (2006) Summary of Global Life Cycle Inventory Data Resources with Contributing TF1 Members: Julie-Ann Chayer and Gyorgyi Cicas, prepared for Task Force 1: Database Registry of the UNEP/SETAC Life Cycle Initiative: 31.

Curran MA, Mann M, Norris G (2005) The International Workshop on Electricity Data for Life Cycle Inventories" *Journal of Cleaner Production* 13:853-862.

Dunmade I (2007) LCA Software Tools and Approaches, Life-Cycle Assessment in Environmental Practice, AGM/Conference, April 25, 2007, The Learning Centre, Mount Royal College, Calgary, Alberta

Ecoinvent database. Swiss Centre for Life Cycle Inventories. http://www.ecoinvent.ch/

EPA (2006) Life Cycle Assessment: Principles and Practice, EPA/600/R-06/060, prepared by Scientific Applications International Corporation for the US Environmental Protection Agency Office of Research and Development, Washington, DC, USA.

Frischknecht R, Althaus H-J, Bauer C, Doka G, Heck T, Jungbluth N, Kellenberger D. and Nemecek T (2007) The environmental relevance of capital goods in life cycle assessments of products and services *Int J Life Cycle Assess.*

Hendrickson C, Lave LB, Matthews HS (2006) *Environmental Life Cycle Assessment of Goods and Services - An Input-Output Approach,* Resources for the Future, Washington, DC.

Heijungs R and Suh S (2002) The Computational Structure of Life Cycle Assessment, Kluwer Academic Publishers, Dordrecht.

ISO (2006) ISO 14044 Environmental management — Life cycle assessment — Requirements and guidelines. International Organization for Standardization, Genève, Switzerland.

Jönbrink AK, Wolf-Wats C, Erixon M, Olsson P, Wallén E (2000) LCA Software Survey, SIK research publication SR 672, IVF research publication 00824, Stockholm, September 2000, www3.ivl.se/rapporter/pdf/B1390.pdf

Koehler A (2008) Water Use in LCA: Managing the Planet's Freshwater Resources *Int J Life Cycle Assess* 13:451-455.

Menke DM, Davis GA, Vigon BW (1996) Evaluation of Life-Cycle Assessment Tools, Battelle 1996, http://isse.utk.edu/ccp/pubs/pdfs/LCAToolsEval.pdf

Quantis (2012) www.quantis-intl.com/software.php, accessed January 16 2012

Ross S, Evans D and Webber M (2002) How LCA Studies Deal with Uncertainty *Int J Life Cycle Assess* 7:47-52

Siegenthaler C,Braunschweig A, Oetterli G, Furter S (2005) LCA Software Guide 2005, Market Overview - Software Portraits, ÖBU, Zürich, Switzerland. www.oebu.ch/fileadmin/media/Publikationen/SR25_LCA_Software_Guide_2005.pdf

Suh S (2009) CEDA 4.0 Users' Guide; http://www.pre-sustainability.com/download/manuals/CEDAUsersGuide.pdf; accessed Janaury 29 2015

Tillman A-M *et al.* (1994) Choice of System Boundaries in Life Cycle Assessment *J Clean Prod* 2(1): p.24.

Unger N, Beigl P, Wassermann G (2004) General requirements for LCA software tools, International Environmental Modelling and Software Society, iEMSs 2004 International Conference, proceedings, www.iemss.org/iemss2004/pdf/infotech/ungegene.pdf, Osnabrück

UNEP/SETAC (2011) Global Guidance Principles for Life Cycle Assessment Databases: A Basis for Greener Processes and Products. Paris, France, United Nations Environment Programme.

Appendix Life Cycle Inventory Data Sources, originally prepared by Joyce Cooper, University of Washington. Available at http://faculty.washington.edu/cooperjs/Definitions/inventory squared.htm (accessed January 2012).

Database Title	Processes, Facility, Sector, or Material-Based	Applicability	Categories of Sectors, Facilities, or Unit Processes	Specific Sectors, Facilities, Unit Processor or Materials Captured	Inputs and Outputs Included
Aerometric Information Retrieval System (US EPA)	Facility	Materials Acqu. & Process, Construct & Manuf., Energy, Transport	Industrial, agricultural, and transportation sources	Example of facilities (sources) include electric power plants, steel mills, factories, and universities	Captures Output Products, Captures Output Wastes (Air emission factors presented per unit of output)
AGRIBALYSE® www.ademe.fr/agribalyse	Agriculture	Crops and animals	Agricultural	More than 100 agricultural products, French as well as a few imported.	
Alternative Fuel & Advanced Vehicles Data Center (US DOE)	Materials	Energy, Transport	Vehicle Fuels	Biodiesel, Electric Fuel, Ethanol. Hydrogen, Methanol, Natural Gas (CNG/LNG), Propane (LPG), P-Series, Solar Fuel	Captures input Energy, Captures Input Materials or land Use, Captures Output Products, Captures Output Wastes (Air emission factors presented per unit of output)

(Continued)

Appendix (Cont.)

Database Title	Processes, Facility, Sector, or Material-Based	Applicability	Categories of Sectors, Facilities, or Unit Processes	Specific Sectors, Facilities, Unit Processor or Materials Captured	Inputs and Outputs Included
Annual Energy Outlook (US DOE Energy Information Administration)	Process and Sector	Energy, Transport	Industrial and other energy use	Residential. Commercial, Industrial, Transportation, Delivered Energy Consumption for all sectors, Electric Generators.	Captures Input Energy, Captures Output Products (energy), Captures Output Wastes
Biennial Reporting System (US EPA)	Facilities	End-of-Life	Treatment, Storage, and Disposal facilities		Captures Output Wastes
Energy Analysis of 108 Industrial Processes by HL Brown (ISBN-10: 0915586932)	Processes	Materials Acqu. & Process, Construct. & Manuf., Energy	Mining, construction, manufacturing, energy production	Meat Packing, Fluid Milk, Canned Fruit and vegetables, fruit and vegetable juices, wet corn milling, bread baking, cakes and pies baking, cane sugar refining, beet sugar refining, soybean oil mills, Malt Beverages, weaving mills, finishing mills, logging camps, sawmills and planning mills, wood products, NEC (Fiberboard), Pulp Mills, paper Board Mills, Corrugated, Solid Fiber Boxes, Building Paper, Alkalies and Chorine, Inorganic Gases, Inorganic Pigments, Industrial inorganic, chemicals, Plastic Materials and Resins, Synthetic Rubbers, Cellulosic Man-made Fibers, organic Fibers, Pharmaceutical Preparations, Cyclic Crudes and Intermediates, Industrial Organic Chemicals, Fertilizers, Chemical	Energy flows

				Preparations, Petroleum Refining, Paving Materials and Blocks, Tires and Inner Tubes, Fabricated Rubber Products, Miscellaneous Plastic Products, Glass, Cement, Brick and Structural Clay Tile, Lime, Gypsum Products, Mineral Wool, Blast Furnaces and Steel Mills, Electrometallurgical Products, Gray Iron Foundries, Primary Copper, Primary Aluminum, Secondary Non-Ferrous Metals, Aluminum Finish Forming, Iron and Steel Forging, Farm Machinery and Equipment, Construction Machinery, Motor Vehicles and Car Bodies, Motor Vehicles Parts and Access, Photographic Film, Photographic Equipment	
E-GRID: Emissions & Generation Resource Integrated Database (US EPA)	Process and Facility	Energy	Energy Production	All US sources - air emissions and kW production by technology type	Captures Output Products, Captures Output Wastes (Air emissions)
IDEMAT (TU Delft, The Netherlands)	Materials	Materials Acqu. & Process, Construct. & Manuf., End-of-Life			Provides Process Flow Diagrams, Captures Input Materials or Land Use

(*Continued*)

Appendix (Cont.)

Database Title	Processes, Facility, Sector, or Material-Based	Applicability	Categories of Sectors, Facilities, or Unit Processes	Specific Sectors, Facilities, Unit Processor or Materials Captured	Inputs and Outputs Included
National Water Use Information Program (USGS Water Resources Division)	Sector	Materials Acqu. & Process, Agriculture, Construct. & Manuf., Use, Energy	Industrial, agricultural, commercial, residential, utilities		Captures Input Materials or land Use (surface and ground water use), Captures Output Wastes (recycling)
Toxics Release Inventory (US EPA)	Facility	Construct. & Manuf.	Manufacturing and government facilities		Captures Input Materials or land Use, Captures Output Products (annual ratio), Captures Output Wastes
US Economic Census (US Census Bureau)	Facility	Materials Acqu. & Process, Construct. & Manuf.	Mining, construction and manufacturing	The Mining sector of the 1997 Economic Census covers all mining establishments of companies with one or more paid employees. Mining is defined as the extraction of naturally occurring mineral solids, such as coal and ores; liquid minerals, such as petroleum, and gases, such as natural gas. The term mining is used in the broad sense to include quarrying, well operations, beneficiating (i.e., crushing, screening, washing, and floatation), and other preparations customarily performed at the mine site or as part of the mining activities (http://www.consus.gov/prod/www/abs/97/ ecmin.html). The construction reports below include	Captures input Energy, Captures Input Materials or land Use, Captures Output Products

				new construction work, additions, alterations, and repairs. Establishments identified as construction management firms are also included. The construction sector is divided into three types of activity or subsectors. The subsectors are the Building, Developing, and General Constructing, Heavy Construction, and Special Trade Contractors. The area reports for the construction industrial contain state and regional level data (http://census.gov/prod/www/abs/97/ecmani.html). Establishments in the manufacturing sector are oftendescribed as plants, factories, or mills and typically use power-driven machines and materials-handling equipment. Also included in the manufacturing sector are some establishments that make products by hand, like custom tailors and the makers of custom draperies; some establishments like bakeries and candy stores that make products on the premises may be included (http://www.census.gov/prod/www/abs/97/ecmani.html)	
Annual and Monthly Energy Reviews (US DOE Energy Information Administration)	Process and Sector	Energy, Transport	Industrial and other energy use	Energy Production by Source, Energy Consumption by source, Energy Consumption by End-Use Sector: household and motor vehicles, Energy Resources: crude Oil and Natural Gas Field Counts; Production, Reserves, Recovery of Oil and Gas, Liquid and Gaseous hydrocarbon, uranium, and petroleum	Captures Input Energy, Captures Output Products, Captures Output Wastes

(Continued)

Appendix (Cont.)

Database Title	**Processes, Facility, Sector, or Material-Based**	**Applicability**	**Categories of Sectors, Facilities, or Unit Processes**	**Specific Sectors, Facilities, Unit Processor or Materials Captured**	**Inputs and Outputs Included**
Annual Estimates of Global Anthropogenic Methane Emissions (Center for Energy and Environmental Studies, Boston Univ.)	Sector	Agriculture, End-of-Life, Energy	Energy, agricultural, and waste storage	Flaring and venting of Natural Gas, Oil and gas Supply Systems, Excluding Flaring, Coal Mining, Biomass Burning, livestock Farming, Rice Farming and Related Activities, Landfills	Captures Output Wastes (methane to air)
EMEP/CORINAIR Atmospheric Emission Factors for Europe (EEA)	Process	Construct. & Manuf.	Industrial sources	Industrial combustion plant and processes with combustion; combustion in boilers, gas; turbines and stationary engines; combustion plants[9]300 MW; combustion plants[9] 50 MW and <300 MW; combustion plants <50 MW Gas turbines; stationary engines; Process furnaces without contact (1); Refinery processes furnaces, coke oven furnaces; blast furnaces cowpers; Plaster furnaces, Processes with contact (2); Sinter plant; Reheating furnaces steel and iron; Gray iron foundries; primary lead production, Primary zinc production, Primary cooper production, Secondary lead production; Secondary zinc production; secondary cooper production,	

				secondary aluminum production, cement, Lime (including iron and steel and paper pulp industries);Asphalt concrete plants; Flat glass; Mineral wool (except binding); Bricks and tiles, Fine ceramic material; Paper mill industry (drying process); alumina production	
Chief Clearinghouse for Inventories and Emission Factors (US EPA)	Process and Facility	Materials Acqu. & Process, Agriculture, Construct. & Manuf., Use, End-of-Life, Energy, Transport	Industrial, agricultural and transportation sources		Captures Output Products, Captures Output Wastes (Air emission factors presented per unit of output)
Database of US Greenhouse Gas Emissions (US DOE Energy Information Administration)	Sector and Materials	Materials Acqu. & Process, Construct. & Manuf., Use, Energy, Transport	Industrial, commercial, residential, transportation, and utilities	Petroleum, coal, geothermal, natural gas, cement production, natural gas, gas flaring, kerosene, jet fuels, transportation (highway vehicles, air transport, vessels), forest fires, LPG, and much more	Captures Output Wastes (greenhouse gases)
Life Cycle Management Of Municipal Solid Waste	Process	End-of-Life	Municipal solid waste management	Collection; Materials Recovery Facility; Energy Model; Transportation; Transfer Stations; Combustion; Compost; Remanufacturing	Provides a Description of Unit Processes, Provides Process Flow Diagrams, Captures Input Energy, Captures Input Materials for land Use, Captures Output Products; Captures Output Wastes

(*Continued*)

Appendix (Cont.)

Database Title	Processes, Facility, Sector, or Material-Based	Applicability	Categories of Sectors, Facilities, or Unit Processes	Specific Sectors, Facilities, Unit Processor or Materials Captured	Inputs and Outputs Included
NAEI: The UK Emission Factors Database (AEA Technology)	Process	Materials Acqu. & Process, Agriculture, Construct. & Manuf., Use, End-of-Life, Energy, Transport	Industrial, agricultural, transportation, and services (hospitals) sources	Mobile Sources: road Traffic, 'Cold Start' emissions, 'hot soak' emissions, rail traffic, airports, ships. Area sources: Emissions from large numbers that are of low significance of small emitters (i.e., domestic gas boilers) or from other identifiable areas (i.e., farmland or landfill sites) are agglomerated together, by type, based on which national grid square they fall in. Point sources: Many of the emissions to the atmosphere resulting from industrial processes and the combustion of fossil fuels are not uniformly spread across urban areas but concentrated at particular points. These point sources include central heating plants serving large groups of buildings, such as hospitals, and boiler plants supplying process heat to industry. They also include industrial processes which require authorization under the Environmental Protection Act 1990.	Captures Output Products, Captures Output Wastes (air emission factors presented per unit of output).
National Agricultural Statistics Service: Agricultural Chemical Use Database (USDA)	Materials	Agriculture	Agriculture	By crop type and by pesticide	Captures Input materials or land Use (pesticide use)

Census of Agriculture (USDA)	Facility and Sector	Materials Acqu. & Process, Agriculture, Construct. & Manuf. Use, End-of-Life, Energy, Transport	Agriculture, pasture, forest, transportation, defense and industrial areas		Captures input Materials or Land use (land use)
Month and State Current Emissions Trends (US DOE Argonne National Laboratory)	Processes	Construct. & Manuf., Energy, Transport	Electric utilities, industrial fuel combustion, commercial/ residential fuel combustion, Industrial processes, transportation, and miscellaneous		Captures Output Wastes (nitrogen oxides, sulfur oxides, and nonmethane VOCs)
National Coastal Pollutant Discharge Inventory (NOAA)	Facility	Materials Acqu. & Process, Agriculture, Construct. & Manuf., Use	Industrial and agricultural emission sources		Capture Output Wastes (water emissions)

(*Continued*)

Appendix (Cont.)

Database Title	Processes, Facility, Sector, or Material-Based	Applicability	Categories of Sectors, Facilities, or Unit Processes	Specific Sectors, Facilities, Unit Processor or Materials Captured	Inputs and Outputs Included
Natural Resources Inventory (USDA)	Sector	Materials Acqu. & Process, Agriculture, Construct. & Manuf., use, End-of-Life, Energy, Transport	Federal and non-Federal lands including agriculture, pasture, forest, etc.		Captures Input Materials or land Use
Recycling Processes (Recycling Data Management Corporation)	Materials	End-of-Life	Industrial and other materials		Captures Input Materials or Land Use, Captures Output Products
RePIS: Renewable Electric Plant Information System (National Renewable Energy Laboratory)	Processes and Facilities	Energy	Electric plants	Geothermal Plants, Hydro Plants, Landfill Methane Plants, Photovoltaic Plants, Solar Thermal Plants, waste to Energy Plants, Wind Plants, Wood and Ag Waste Plants	Captures Output Products (energy)
Renewable Resource Data Center	Sector	Energy	Energy efficiency and renewable energy		Provides a Description of Unit Processes, Captures Input

(National Renewable Energy Laboratory)					Energy, Captures Input Materials or Land Use, Captures Output Products, Captures Output Wastes
Engineered Materials Abstracts	Processes and Materials	Materials Acqu. & Process, Agriculture, construct. & Manuf., use, End-of-Life, Energy, Transport	Industry, utilities, agriculture, services, etc.		Provides a Description of Unit processes, Captures Input Energy, Captures Input Materials or land Use, Captures Output Products, Captures Output Wastes
Kirk-Othmer *Encyclopedia of Chemical Technology*	Process, Materials	Materials Acqu. & Process, Construct. & Manuf., End-of-Life, Energy, Transport	Chemical processes	Areas of chemical technology that will deal with industrial products, natural materials, and processes in such fields as: agricultural chemicals, chemical engineering, coatings and inks, composite materials, cosmetic and pharmaceuticals, dyes, pigments and brighteners, ecology and industrial hygiene, energy conservation and technology, fats and waxes, fermentation and enzyme technology, fibers, textiles and leather, food and animal nutrition, fossil fuels and derivatives, glass, ceramics and cement, industrial inorganic chemicals, industrial organic chemicals, metals, metallurgy and metal alloys, plastics and elastomers, semiconductors and emulsion technology, water supply, purification and reuse, wood, paper, and industrial carbohydrates. Also	Provides a Description of Unit Processes, Provides Process Flow Diagrams, Captures Input Energy, Captures Input Materials or Land Use, Captures Output Products, Captures Output Wastes

(*Continued*)

Appendix (Cont.)

Database Title	Processes, Facility, Sector, or Material-Based	Applicability	Categories of Sectors, Facilities, or Unit Processes	Specific Sectors, Facilities, Unit Processor or Materials Captured	Inputs and Outputs Included
				includes miscellaneous topics: instrumentation and quality control, information retrieval, maintenance, market research, material allocation and supply, legal issues, process development and design, product development and technical service, research and operations management (systems management, networks, etc.), and transportation of chemical products.	
Long Term World Oil Supply (A Resource Base/ Production Path Analysis (US DOE Energy Information Administration)	Process	Energy	Oil Production		Provides a Description of Unit Processes, Captures Output Products (oil)
Mineral Industry Survey (USGS)	Materials	Materials Acqu. & Process	Minerals Production		Captures Input Materials or land use, Captures Output Products, Captures Output Wastes (recycling)
Municipal Solid Waste Factbook (US EPA)	Facilities	End-of-Life	Household waste management practices		Provides a Description of Unit Processes

Municipal Solid Waste Survey (US EPA Office of Solid Waste)	Materials	End-of-Life	Municipalities		Captures Input Materials or land Use (materials into landfills and incineration), captures Output Wastes (recycling)
Net Generation and Utility Retail Sales (US Census Bureau)	Process	Energy	Energy Production	Electric utilities; fossil fuels (primarily coal), nuclear, renewable resources.	Captures Output Products
Search US Patents	Processes and Materials	Materials Acqu. & Process, Agriculture, construct. & Manuf., Use, End-of-Life, energy, Transport	Industry, utilities, agriculture, services, etc.		Provides a Description of Unit Processes, Captures Input Energy, Captures Input Materials or land Use, Captures Output Products, Captures Output Wastes

(Continued)

Appendix (Cont.)

Database Title	Processes, Facility, Sector, or Material-Based	Applicability	Categories of Sectors, Facilities, or Unit Processes	Specific Sectors, Facilities, Unit Processor or Materials Captured	Inputs and Outputs Included
Sector Notebooks (US EPA)	Process	Materials Acqu. & Process, Agriculture, Construct. & Manuf., Energy	Manufacturing and Agriculture	Agricultural Chemical, Pesticide and Fertilizer Industry (1999), Agricultural Crop Production Industry (1999); Agricultural Livestock Production Industry (1999); Aerospace Industry (1998); (new); Air Transportation Industry (1997); Dry Cleaning Industry (1995); Electronics and Computer Industry (1995); Fossil Fuel Electric Power Generation Industry (1997); inorganic Chemical Industry (1995); Iron and Steel Industry (1995); Lumber and Wood Products Industry (1995); Metal Casting Industry (1997); Metal Fabrication Industry (1995); Metal Mining Industry (1995); Motor Vehicle Assembly Industry (1995); Nonferrous Metals Industry (1995); Non-Fuel, non-Metal Mining Industry (1995); Oil and Gas Extraction Industry (1999); (new); Organic Chemical Industry (1995); Petroleum Refining Industry (1995); Pharmaceutical Industry (1997); Plastic Resins and Man-made Fibers Industry (1997); Printing Industry (1995); Pulp and paper Industry (1995); Rubber and Plastic Industry (1995); Shipbuilding and Repair Industry (1997); Stone, Clay, Glass and Concrete Industry (1995); Textiles Industry (1997); Transportation Equipment Cleaning industry (1995); Wood Furniture and Fixtures Industry (1995)	Provides a Description of Unit Processes, Provides Process Flow Diagrams, Captures Input Energy, Captures Input Materials or Land Use, Captures Output Products, Captures output Wastes

Ullmann's *Encyclopedia Of Industrial Chemistry*	Process, Materials	Materials Aquc. & Process, Agriculture, Construct. & Manuf., Use, End-of-Life, Energy, Transport	Chemical Processes	Over 800 articles written by 3000 experts, more than 10,000 tables, 20,000 figures, 16 million words - whatever way you look at it, Ullmann's Encyclopedia of Industrial Chemistry offers you a stupendous amount of information in industrial chemistry, process engineering, materials science, environmental chemistry, food science and biotechnology	Provides a Description of Unit Processes, Provides Process Flow Diagrams, Captures Input Energy, Captures Input Materials or Land use, Captures Output Products, Captures Output Wastes
Software and Tools Within Chief Clearinghouse for Inventories And Emission Factors (US EPA)	Process and Facility	Materials Acqu. & Process, Agriculture, Construct. & Manuf., use, End-of-Life, Energy, Transport	Industrial, agricultural and transportation sources		Captures Output Products, Captures Output Wastes (Air emission factors presented per unit or output)

(*Continued*)

Appendix (Cont.)

Database Title	**Processes, Facility, Sector, or Material-Based**	**Applicability**	**Categories of Sectors, Facilities, or Unit Processes**	**Specific Sectors, Facilities, Unit Processor or Materials Captured**	**Inputs and Outputs Included**
The Greenhouse Gases, Regulated Emissions, and Energy Using Transportation (GREET) Model (Argonne National Laboratory)	Process	Energy, Transport	Transportation and fuel cycle emissions	Gasoline vehicles; Federal reformulated gasoline, California reformulated gasoline, ESO; CIDI vehicles: diesel); compressed natural gas vehicles; bi-fuel, dedicated fuel; Dedicated Liquefied petroleum gas vehicles, Flexible-fuel vehicles, E85, M85; Electric vehicles; Grid-connected HEVs. California reformulated gasoline; Grid-independent HEVs; Federal reformulated gasoline, diesel S1 vehicles; Dedicated compressed natural gas, dedicated liquefied natural gas; dedicated liquefied petroleum gas, dedicated E90, dedicated M90; CIDI vehicles; Federal reformulated gasoline, California reformulated diesel, dimethyl ether, Fischer-Tropsch diesel; biodiesel, Grid-independent HEVs. Federal reformulated gasoline, compressed natural gas, liquefied natural gas, liquefied petroleum gas, E90, M90, reformulated diesel, dimethyl ether, fischer-Tropsch diesel, biodiesel; Grid-connected HEVs. California reformulated gasoline, compressed natural gas, liquefied natural gas, liquefied petroleum gas, E90, M90, reformulated diesel, dimethyl ether, Fischer-Tropsch diesel, biodiesel; Electric vehicles, Fuel-cell vehicles; Hydrogen, methanol. Gasoline, ethanol, compressed natural gas	Captures Input Energy, Captures Input Materials or land Use, Captures Output Products, Captures Output Wastes

MOBILE6: Vehicle and Engine Emission Modeling Software (US EPA)	Process	Transport	Transportation		Captures Output Products, Captures Output Wastes (Air emission factors presented per unit of output)
National Emissions Inventory (NEI) Air Pollutant Emissions Trends Data (US EPA)	Individual point source, facility, county level	Emission estimates for area, mobile and other sources		Information from numerous State and local air agencies, tribes, and industry on stationary and mobile sources that emit criteria air pollutants and their precursors, as well as hazardous air pollutants (HAPs).	Estimates of annual emissions, by source, of air pollutants in each area of the country, on an annual basis.

Chapter 3 Exercises

1. Data Sources
   A life cycle inventory is compiled from various databases and data sources for the necessary background and foreground data. Name examples of data sources that could be useful for providing background unit process data. How might foreground data be collected?
2. Co-Product Allocation
   An industrial process produces two co-products: 850 kg of product A and 430 kg of product B. A is sold at $2.00 per kg. B is sold for $0.30 per kg.
   Discuss how the inventory data for Product A would be modeled according to the ISO hierarchy following:
   a) Subprocess modeling
   b) System Expansion
   c) Physical Property Allocation (for example, by weight)
   d) Economic Basis (market value)
3. Avoided Burden Credit
   Some studies allow for what is called a "displacement credit" to account for impacts that are declared to be avoided because of additional functions provided by the system being studied. Provide an example of a waste stream or a co-product which could be viewed as displacing a competing product. What data could be used to determine the basis of the reference flow to account for the amount of displaced product? Explain when an avoided burden approach has merit and when it could be used to skew the results?
4. System Expansion for Co-Product Allocation and Avoided Burden
   What is the difference between a system expansion approach for use in co-product allocation versus use in calculating avoided burdens?
5. What is the difference between closed loop and open loop recycling? How does each one affect modeling choices?
6. Cut-Off Rule
   a) What does it mean to use the cut-off rule for modeling a recycled
   b) What is a reasonable cut-off threshold for input materials (on a weight basis)?

4

Life Cycle Impact Assessment

Abstract
The goal of life cycle impact assessment is to convert the large quantity of input/output data generated in the life cycle inventory analysis phase of an ISO Life Cycle Assessment (LCA) into relevant impact category indicators. The ISO 14044 standard establishes both requirements and recommendations for the choice of impact categories, category indicators and characterization models to be used in LCIA as part of an LCA study. This selection process must be done at the outset of a study, during the goal and scope definition phase. Conducting LCIA has been greatly simplified by the increasing availability of software programs. Choosing which one to use is challenging, requiring the practitioner to understand the main characteristics of these methods and keep up-to-date with the latest developments. While the chapter does not attempt to explain in great detail how the individual models within LCIA programs operate, or what the underlying assumptions are, it describes the many currently available LCIA methods and indicators.

References from the LCA Handbook

Aims of the Chapter

1. Outline the steps involved in conducting life cycle impact assessment according to ISO 14040.
2. Present the various LCIA models, both midpoint and endpoint, that are available and how they differ.
3. Provide practical guidance by describing the key characteristics of current LCIA methods available in LCA software.

4.1 Introduction

The life cycle impact assessment (LCIA) phase of an LCA aims to help users understand and evaluate the magnitude and significance of the potential environmental impacts for a product system throughout the life cycle of the product. Its motivation comes from two observations:

- The final result of the inventory analysis, the inventory table, is very long (e.g., 1000 different items) and difficult to handle;
- The inventory table contains many items that require expert knowledge (such as 2-methyl-2-butene) to understand their importance.

LCIA methodology does not necessarily attempt to quantify site- specific or actual impacts associated with a product, process, or activity. Rather, the individual impact methods convert LCI results to common units and aggregate the converted results within the same impact category. By modeling possible impact pathways, LCIA addresses ecological and human health effects, as well as resource depletion, in order to link the product or process being studies and its potential environmental impacts.

While the unit process is the central element of the inventory analysis, the central element in impact assessment is the impact category. ISO defines it as a "class representing environmental issues of concern to which life cycle inventory analysis results may be assigned." Perhaps more helpful are some examples: climate change, toxicity, and depletion of fossil energy carriers.

Since climate change (often used interchangeable with the term "global warming") is a well-known issue, it is useful for illustrating the main ideas of impact assessment. The inventory table contains a number of greenhouse gases: CO_2, CH_4, N_2O, etc. These are known to contribute all to the phenomenon of climate change. Climate change involves long sequence of causal mechanisms: emissions of greenhouse gases lead to changes in the composition of the atmosphere, which lead to a change in the radiation balance, which in turn leads to a change in the temperature distribution, which leads to changes in climate, which leads to changes in ecosystems and human activities, etc. The further we proceed in this causal chain, the more uncertain and speculative our knowledge becomes. While quite some scientific evidence is available with respect to the composition of the atmosphere, the impacts on biodiversity are debated. Many of these later impacts are even conditional on our future activities, including future emission scenarios and mitigating actions. To be able to quantitatively model the emissions

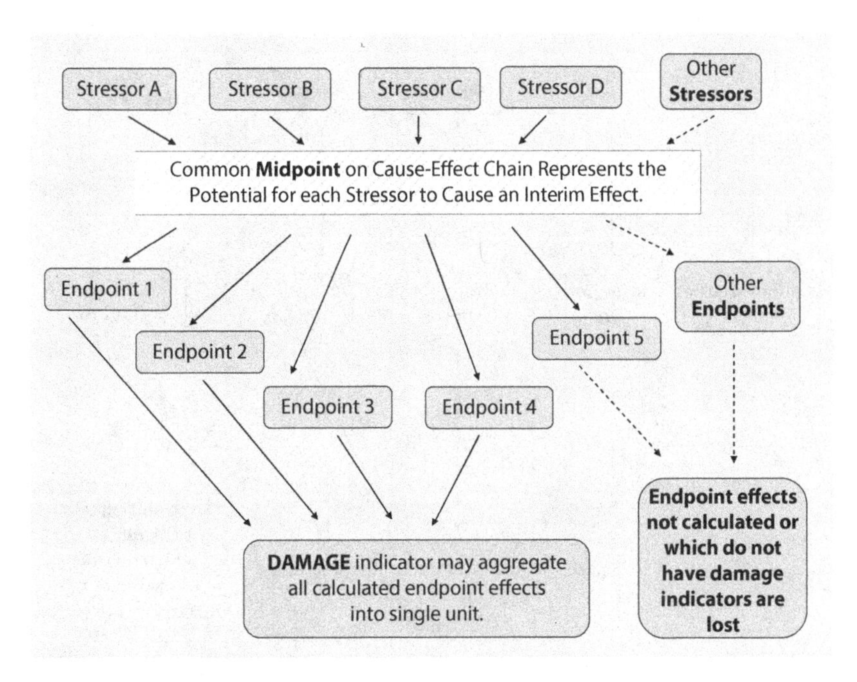

Figure 4.1 Impact models progresses from inventory flow to damage for classic midpoint impact categories (Bare and Gloria 2006).

of different greenhouse gases into an impact indicator for climate change, we must do several things.

First, we must choose a certain point in the causal mechanism. This can be at the front end (change in radiation balance), at the back-end (change of biodiversity), or somewhere in between (change in temperature). In LCA, two main schools have emerged:

- Those that focus on the midpoint approach;
- Those that focus on the endpoint approach.

The midpoint approach has the advantage that it includes fewer debatable assumptions and less-established facts; the endpoint approach has the advantage that it provides more intuitive metrics (like loss of life years instead of kg CO_2-equivalents). Regardless of the choice between midpoint and endpoint, the indicator chosen is referred to as the impact category indicator, or category indicator for short.

Second, a way must be found to convert the emission data into the chosen impact indicator. Scientists in chemistry, meteorology, ecology, etc, have developed model fragments to estimate the atmospheric life-times of greenhouse gases, their effect on the radiation balance and the formation of clouds, the effects of temperature on the distribution of species, etc. These fragments have been combined by workgroups from the UN-based International Panel on Climate Change (IPCC) into quantitative models of the impacts of greenhouse gas emissions. Part of this is the global warming potentials (GWPs), which are quantitative measures of the strength of different greenhouse gases. Many midpoint LCIA methods apply GWPs for climate change. This type of characterization calculation is applied for each impact indicator category (see Figure 4.3).

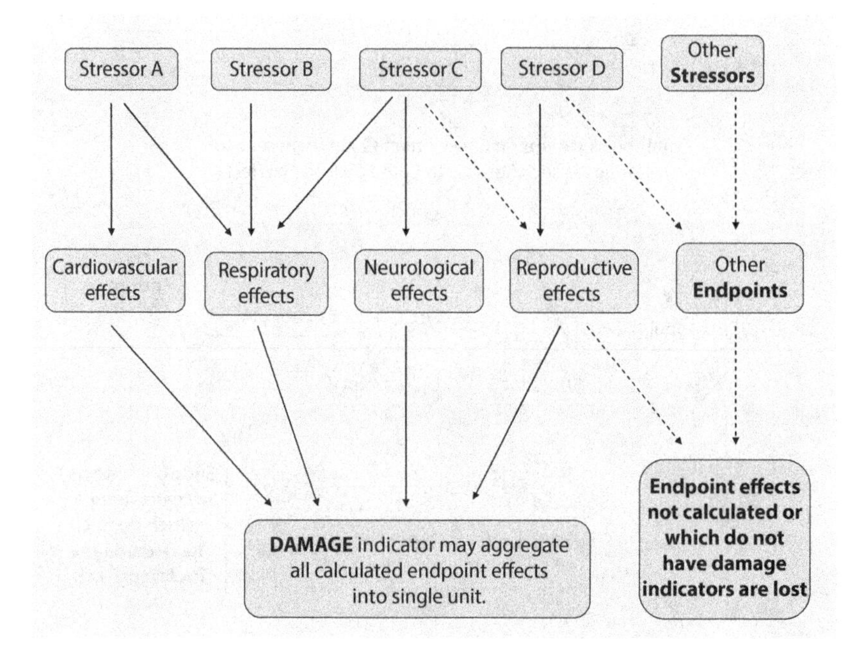

Figure 4.2 Progression from inventory flows to damage for human health (Bare and Gloria 2006).

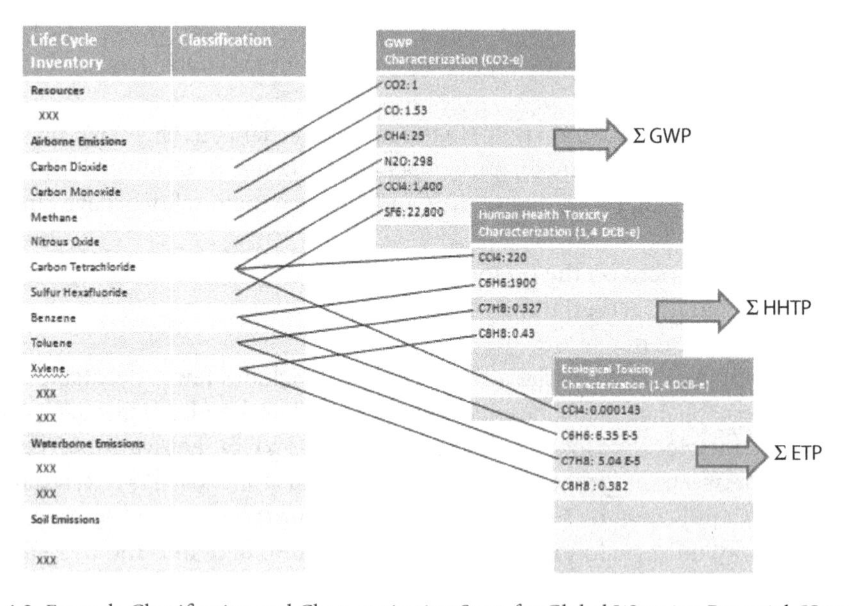

Figure 4.3 Example Classification and Characterization Steps for Global Warming Potential, Human Health Toxicity and Ecological Toxicity (from Industrial Ecology Consultants 2010).

As a concrete example of how characterization works, let us study a fragment of a hypothetical inventory table, containing the following information: emission of carbon dioxide (CO_2) 100 kg, emission of methane (CH_4) 10 kg, emission of sulfur dioxide (SO_2) 1 kg. Characterizing greenhouse gases with GWPs requires a table with GWPs. In such a table, one can find that the GWP of CO_2 is 1 (by definition) and that the GWP of CH_4 is 25 (kg CO_2-equivalent/kg CH_4). SO_2 has no GWP; it is assumed not to contribute to climate change. Characterization now proceeds in the case of climate change by calculating

$$(1x100) + (25x10) = 350\ kg_{CO_2\text{-equivalents}}$$

For the more general case, this can be written as

$$GW = \sum_s GWP_s \times m_s$$

where GW is the global warming score, s the substance (the different greenhouse gases), GWP_s the GWP of substance s, and m_s the emitted amount of substance s in kg. This may be further generalized as

$$I_c = \sum_s CF_{c,s} \times m_s$$

where c stands for the impact category, I represents the indicator result for category c, and $CF_{c,s}$ is the characterization factor that links substance s to impact category c. This formula is the operational formula for characterization. With a table of characterization factors specified, it makes clear that:

- LCIA builds on the results of LCI (as is clear from the term m_s);
- Characterization converts the results of LCI into a common metric (as is clear from the multiplication by CF);
- Characterization aggregates the converted LCI results (as is clear from the summation symbol).

Characterization results in a list of numbers, referred to as "scores" (for instance a score for climate change, a score for toxicity, etc.) ISO refers to such numbers as "category indicator results," but most LCA practitioners prefer the name "score," sometimes expanded with the name of the impact (as in "toxicity score"). The complete list is known by names such as "LCIA profile," "characterization table," etc.

Even though the remainder of the environmental mechanism from midpoint to endpoint to damages describes the link to environmentally relevant endpoint indicators, this sometimes occurs at the expense of the comprehensive nature of the midpoint, and likely resulting in higher uncertainty. In certain categories, providing methodological approaches that characterize the environmental mechanism closer to endpoints and damages does not provide additional distinction of differences in impact between substances. However, a model between damage and midpoint may add relevance (either in a quantitative or qualitative manner – in cases where quantification of endpoints is difficult to impossible), and this relevance may be added for all substances in the same way. This could also enable us to compare the outcomes of different midpoint categories using models based on natural science instead of weighting factors based on social science. In a midpoint model it seems wise to minimize the unnecessary uncertainty by choosing a midpoint indicator as early as possible in the environmental chain where all substances are unified in an indicator yet the five criteria are still satisfied: comprehensiveness, relevance/reproducibility, transparency, validity and compatibility.

The second group of impact categories, illustrated in Figure 4.2, may not have a common midpoint and are comprised of different environmental mechanisms. Examples of the second type of impact categories which are almost always represented at an aggregated level (either at damage or midpoint level) include human toxicity and ecotoxicity, where interim human health endpoints that may be aggregated include neurological, reproductive, respiratory, and cardiovascular health endpoints. The aggregation may be in units of DALYs, monetary value, or a unitless score which is based on the relative human toxicity potency after including the fate, transport, and toxicity of the substances and comparing to a reference substance.

The ILCD Handbook (JRC 2010) suggests considering the following points:

1. For the first group of impact categories described above, the goal of damage modeling is to make results in different midpoint categories comparable, and sometimes to arrive to a single score, or smaller number of environmental scores. It can then replace or support weighting practices in the midpoint approaches. The choice to stay at the midpoint level or go to the damage level is left to the user.
2. When the decision has been made to go to the damage level on an impact category of the first type (e.g., global climate change), care must be taken to ensure comprehensiveness. For example, while it may be relatively easy to quantify some impacts (e.g., malaria), other impacts (e.g., the impact on biodiversity) may not be so easily quantified and thus may be lost.
3. Intermediary steps should be made explicit and reported separately. For example, if number of cases, Years of Life Lost (YLL) and Years of Life Disabled (YLD) are utilized then these should be considered separately first for impacts on human health. Disability weighting could then be explicitly considered if desired to group diseases together to arrive at DALY.
4. All modeling (midpoint and damage) should be properly documented on data and modeling uncertainty and reliability. Value choices should be made explicit and properly documented. As a matter of fact, it is important to be more specific about these values choices to decrease the uncertainty. There is no unique universal set of values.

In the end, LCIA approaches are typically viewed along one of two families: classical methods that determine impact category indicators at an intermediate position of the various impact pathways (e.g. ozone depletion potential) or damage-oriented methods that aim to present results in the form of damage indicators at the level of an ultimate societal concern (e.g. harm to human health).

4.2 Choice of Impact Models and Categories

GWPs provide one example of a set of characterization factors, and that the IPCC-model from which they are derived is an example of a characterization model. Note, by the way, that IPCC has not developed this model as a characterization model for LCIA, but that the LCA-community has adopted this model as such and its derived GWPs as characterization factors. Also note that the characterization model itself is not used by LCA practitioners; only the characterization factors that have been derived from it as a one-time exercise are used. Characterization factors are often tabulated in LCA guidebooks and are implemented in many LCA software packages, while the characterization models often require supercomputers and expert knowledge.

In fact, one element is needed before one can select a category indicator and a characterization model with associated characterization factors. It is the selection of impact categories to be addressed. Some LCA studies concentrate on just one impact category. For instance, the carbon footprint (of a product, not of a company or country) is considered a form of LCA that addresses just climate change at the midpoint level through GWPs. At the other extreme, some LCA studies incorporate fifteen or more impact categories. For consistency reasons, the choice of impact categories is often made on the basis of a recommended impact assessment guidebook or its implementation in software. Thus, in practice one often sees LCA-studies reporting the use of "IMPACT2002+," "TRACI," "CML-IA," "ReCiPe," "ILCD," etc. All these methods comprise a recommended set of impact categories with a category indicator and set of characterization factors. ISO does not specify any choice in these matters. Table 4.1 gives an overview of some often-used impact categories and category indicators. We see that the column with endpoint indicators contains many times the same term (e.g., "loss of life years"). This suggests that impact categories can be aggregated into fewer endpoint indicators than midpoint indicators.

The goal in assigning LCI results to the impact indicator categories is to highlight environmental issues associated with each. Assignment of LCI results should:

- attempt to assign results which are exclusive to an impact category and
- identify LCI results that relate to more than one impact category, including distinguishing between parallel mechanisms such as SO_x allocated between human health and acidification, and allocating among serial mechanisms such as NO_x assigned to smog formation and acidification.

Figure 4.4 depicts the mapping of environmental releases (interventions) quantified in an LCI to the appropriate midpoint categories. Note how some inventory releases have more than one arrow directed to a midpoint. Similarly, one midpoint impact category may contribute to more than one endpoint impact.

Typically in impact assessment, a "non-threshold" assumption is used. That is, inventory releases are modeled for their potential impact regardless of the total load to the receiving environment from all sources or consideration of the assimilation capacity of the environment.

If LCI results are unavailable, or of insufficient quality to achieve the goal of the study, then either an iterative data collection effort or an adjustment of the goal is required.

4.3 Current LCIA Approaches

The following sections describe current approaches that are being applied to model some of the impact category indicators listed in Table 4.1. The most simplistic models

Table 4.1 Common Impact Categories with Example Midpoint and Endpoint Indicators.

Impact Category	Midpoint Category Indicator	Endpoint Category Indicator
climate change	infra-red radiative forcing	loss of life years, fraction of disappeared species
ozone layer depletion	change in tropospheric ozone concentration	loss of life years
acidification	H+ concentration	fraction of disappeared species
eutrophication	biomasss potential	fraction of disappeared species
human toxicity (sometimes split into carcinogenics, non-carcinogenics, respiratory effects, etc.)	time-integrated exposure, corrected for hazard	loss of life years
eco-toxicity (sometimes split into aquatic toxicity, terrestrial toxicity, marine toxicity, etc.)	time-integrated exposure, corrected for hazard	fraction of disappeared species
depletion of energy carriers	primary energy requirement	decreased availability
depletion of material resources	amount of material used, corrected for availability and/or importance	decreased availability
land use impacts	amount of land occupied or transformed	fraction of disappeared species
water use impacts	amount of water used or displaced	decreased availability

are described in order to offer insight into the types of approaches that are being considered useful from both a practical aspect as well as least cost.

4.3.1 Stratospheric Ozone Depletion

Ozone depletion is suspected to be the result for the release of man-made halocarbons, e.g., chlorofluorocarbons (CFC)'s, that migrate to the stratosphere. For a substance to be considered as contributing to ozone depletion, it must:

- be a gas at normal atmospheric temperatures,
- contain chlorine or bromine, and
- be stable within the atmosphere for several years. (Wenzel *et al* 1997)

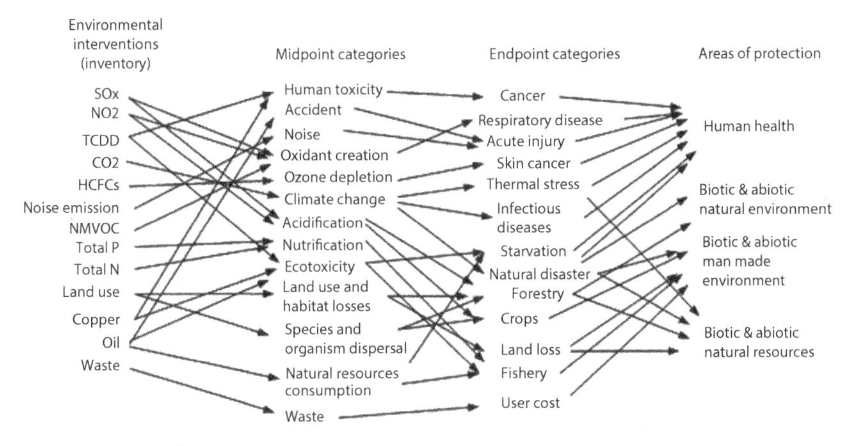

Figure 4.4 Assigning inventory releases to midpoint categories, and midpoint to endpoint categories (from Jolliet *et al* 2004).

The most important groups of Ozone Depleting Compounds (ODC's) are the CFC's, HCFC's (hydrochlorofluorocarbons), halons and methyl bromide. HFC's (hydroflouro-carbons) are also halocarbons but contain fluorine instead of chlorine or bromine, and are therefore not regarded as contributors to ozone depletion.

Ozone depletion potential (ODP) is calculated by multiplying the amount of the emission (Q) by the equivalency factor:

$$\mathrm{ODP} = \mathrm{Q} \bullet \mathrm{EF}$$

The current status on reporting Equivalency factors is to use CFC11 as the reference substance. The Equivalency Factor is defined as:

$$\mathrm{EF_{ODP}} = \frac{\text{Contribution to stratospheric ozone depletion from n over \# years}}{\text{Contribution to stratospheric ozone depletion from CFC11 over \# years}}$$

General LCA practice uses values that represent ODC's full contribution, but Table 4.2 also gives factors for 5, 20 and 100 years for some gases. These individual potentials can then be summed to give an indication of projected total ODP, for substances 1 through *n* in the life cycle inventory that contribute to ozone depletion:

$$\text{Total ODP} = \sum_{1}^{n} \left(Q_i \times \mathrm{ODP}_i\right)$$

Where Q_i is the mass of ozone-depleting gas *i* assigned to the category.

4.3.2 Global Warming Potential

The most significant impact on global warming has been attributed to the burning of fossil fuels, such as coal, oil and natural gas. Several compounds, such as carbon dioxide (CO_2), nitrogen dioxide (N_2O), methane (CH_4) and halocarbons, have been identified as substances that accumulate in the atmosphere leading to an increased global warming effect.

Table 4.2 Equivalency Factors for Ozone Depletion Potential (Wenzel *et al* 1997).

Substance	Formula	ODP g CFC11/g substance			
		5 years	20 years	100 years	∞
CFC11	$CFCl_3$	1	1	1	1
CFC12	CF_2Cl_2	0.55	0.59	0.78	0.82
CFC113	$CF_2ClCFCl_2$	1.26	1.23	1.14	0.90
CFC114	CF_2ClCF_2Cl				0.85
CFC115	CF_2ClCF_3				0.40
Tetrachloromethane	CCl_4				1.20
HCFC22	CHF_2Cl	0.19	0.14	0.07	0.04
HCFC123	CF_3CHCl_2	0.54	0.33	0.13	0.014
HCFC124	CF_3CHFCl	0.17	0.14	0.08	0.03
HCFC141b	$CFCl_2CH_3$	1.03	0.45	0.15	0.10
HCFC142b	CF_2ClCH_3				0.05
HCFC225ca	$CF_3CF_2CHCl_2$				0.02
HCFC225cb	CF_2ClCF_2CHFCl				0.02
1,1,1-trichlorethane	CH_3CCl_3				0.12
Methyl chloride	CH_3Cl				0.02
Halon1301	CF_3Br	10.3	10.5	11.5	12
Halon 1211	CF_2ClBr	11.3	9.0	4.9	5.1
Methyl bromide	CH_3Br	15.3	2.3	0.69	0.64

For a substance to be regarded as a global warmer, it must:

- be a gas at normal atmospheric temperatures, and
- either be able to absorb infrared radiation and be stable in the atmosphere with a long residence time (in years) OR
- be of fossil origin and converted to CO_2 in the atmosphere. (Wenzel *et al* 1997)

Table 4.3 is a suggested list of substances that are considered to contribute to global warming. Equivalency Factors, based on carbon dioxide as 1, are shown for each substance over 20, 100 and 500 year spans. The choice of timescale can have considerable effect on how global warming potential is calculated. The 100 year time frame is often selected, unless some reason exists that indicates otherwise.

The Global Warming Potential (GWP) is calculated by multiplying a substance's mass emission (Q) by its equivalency factor. These individual potentials can then be summed to give an indication of projected total GWP, for substances 1 through n in the life cycle inventory that contribute to global warming:

$$\text{Total GWP} = \sum (Q_i \times GWP_i)$$

Where Q_i is the mass of global warming gas *i* assigned to the category.

Table 4.3 Equivalency Factors for Global Warming (Wenzel *et al* 1997).

Substance	Formula	GWP $g\ CO_2/g$ substance		
		20 years	100 years	500 years
Carbon dioxide	CO_2	1	1	1
Methane	CH_4	62	25	8
Nitrous oxide	N_2O	290	320	180
CFC11	$CFCl_3$	5000	4000	1400
CFC12	CF_2Cl_2	7900	8500	4200
CFC113	$CF_2ClCFCl_2$	5000	5000	2300
CFC114	CF_2ClCF_2Cl	6900	9300	8300
CFC115	CF_2ClCF_3	6200	9300	13000
Tetrachloromethane	CCl_4	2000	1400	500
HCFC22	CHF_2Cl	4300	1700	520
HCFC123	CF_3CHCl_2	300	93	29
HCFC124	CF_3CHFCl	1500	480	150
HCFC141b	$CFCl_2CH_3$	1800	630	200
HCFC142b	CF_2ClCH_3	4200	2000	630
HCFC225ca	$CF_3CF_2CHCl_2$	550	170	52
HCFC225cb	CF_2ClCF_2CHFCl	1700	530	170
1,1,1-trichloroethane	CH_3CCl_3	360	110	35
Chloroform	CH_3Cl	15	5	1
Methylene chloride	CH_2Cl_2	28	9	3
HFC 134a	CH_2FCF_3	3300	1300	420
HFC152a	CHF_2CH_3	460	140	44
Halon 1301	CF_3Br	6200	5600	2200
Carbon monoxide*	CO	2	2	2
Hydrocarbons (NMHC)*	various	3	3	3
Partly oxidized hydrocarbons*	various	2	2	2
Partly halogenated hydrocarbons*	various	1	1	1

* contributes indirectly due to conversion into CO_2. Only compounds of petrochemical origin.

4.3.3 Nonrenewable Resource Depletion Potential

This impact category models resources that are nonrenewable, or depletable. The sub-categories include:

- Fossil fuels
- Net Non-Fuel Oil and Gas
- Net Mineral Resources, and
- Net Metal Resources.

Some models also include the energy that is inherent in a product that is made from a petroleum feedstock in order to reflect the amount of stock that was diverted and is no longer available for use as an energy source.

This category can also reflect land use as a resource. Land that has been disturbed directly due to physical or mechanical disturbance can be accounted for as a resource that is no longer available either from human use or for ecological benefit (such as providing habitat for a certain species).

Nonrenewable resource depletion is a frequently discussed impact category with no consensus on how to model it. While different methodologies being applied, most are based on calculated world reserves, extraction rates, recycling, waste generation and natural accretion.

The reserve Base-to-Use Ratio can be calculated as follows:

$$\frac{\text{ReserveBase}(R)}{\text{Use}(U)} = \text{Number of years of remaining use left}(\text{at current use rate})$$

$$\frac{\text{Use}(U)}{\text{Reserve Base}(R)} = \%\text{ of reserve base used}$$

The recycled resource is linked to the original virgin material use and corresponding reserve base. Emissions are not spatially or temporally lined to the original virgin unit operation.

$$\text{Accounting for all reserve bases}: \frac{\text{Waste}(\Sigma W)}{\text{Reserve base}(R) + \text{Recyclable Stoke}(\Sigma S)}$$

The current assumption is that only one iteration of recycling and material integrity is sustained. If natural accretion is accounted for, the following formula results:

$$\frac{\text{Waste}(\Sigma W) - \text{Natural Accretion}(N)}{\text{Reserve Base}(R) + \text{Subsequent Uses}(\Sigma S)}$$

Including time period in the equation, we get:

$$\frac{(\Sigma W - N)\Delta T}{R + (\Sigma S)\Delta T}$$

Current assumption: $\Delta T = 50$ years

And accounting for baseline reserve bases is:

$$\frac{(\Sigma W - N)\Delta T + (R_b - R)}{R + (\Sigma S)\Delta T}$$

Rb = A reserve base baseline

$$\therefore \text{Resource Depletion Factor}(\text{RDF}) = \frac{(\Sigma \text{W} - \text{N})\Delta\text{T} + (\text{R}_\text{b} - \text{R})}{\text{R} + (\Sigma \text{S})\Delta\text{T}}$$

Resource depletion of fossil fuels represents a simple application. Accretion is zero and recycling is nil. Thus, wasted resource equals resource used, or:

$$\text{RDF} = \frac{(\text{W}) * \text{T}}{\text{R}}$$

The impact for the resource depletion category can then calculated according to the formula:

Resource Depletion Indicator (RD) = Resource Use x Resource Depletion Factor (RDF)

For net resources depleted (or accreted), the units of measure express the equivalent depletion (or accretion) of the identified resource. All of the net resource calculations are based on RDF's.

4.3.4 Acidification Potential

For acidification an equivalency approach is typically applied and the stressor flows are converted into SO_2 equivalents. For example, NO_2 is multiplied by 64/(2*46) = 0.70 since this is the molar proton release potency of NO_2 compared to SO_2. Table 4.5 shows sample calculations using potency factors for an inventory with SO_2, NO_2, and HCl releases. This approach takes the calculation one step further and includes an emission loading factor to reflect how much of the inventory release is expected to reach the receiving environment (emission loading).

Table 4.4 Units of Measure for Resource Indicators.

Indicator - Net Resource	Units of Measure
Water	equivalent cubic meters
Wood	equivalent cubic meters
Fossil Fuels	tons of oil equivalents
Non-Fuel Oil and Gas	tons of oil equivalents
Metals	tons of (metal) equivalents
Minerals	tons of (mineral) equivalents
Land Area	equivalent hectares

Table 4.5 Calculating Acidification "Emission Loading" (SCS 2010).

Unit Operation	Inventory Emission	LCI Result (ton/30a)	Potency Factor	Molar Equivalent (ton/30a)	Character-ization Factor	Emission Loading (ton/30a)
Coal mining/ transport	SO_2	31620	1	31620	0.5	15810
	NO_2	9660	0.7	6762	0.3	2029
	HCl	270	0.88	238	0.5	119
CaO product/ transport	SO_2	240	1	240	0.15	36
	NO_2	1260	0.7	882	0.075	66
Coal use	SO_2	50190	1	50190	0.15	7529
	NO_2	36480	0.7	25536	0.075	1915
	HCl	5210	0.88	13385	0.15	2008
Total				**128853**		**29512**

4.3.5 Eutrophication Potential

Eutrophication occurs in aquatic systems when the limiting nutrient in the water is supplied, thus causing algal blooms. In fresh water, it is generally phosphate which is the limiting nutrient, while in salt waters it is generally nitrogen which is limiting. In general, addition of nitrogen alone to fresh waters will not cause algal growth, and addition of phosphate alone to salt waters will not cause significant effects. In brackish waters, either nutrient can cause algal growth, depending on the local conditions at the time of the emissions.

Eutrophication is generally measured using the concentration of chlorophyll-a in the water. Waters with less than two milligrams chlorophylla per cubic meter (2 mg chla m^3) are considered "oligotrophic," while those with 2 to 10 mg chla m^3 are considered "mesotrophic," and those greater than 10 mg chla m^3 are termed "eutrophic." Over 20 mg chla m^3 waters are considered "hypereutrophic."

As waters become mesotrophic, their species assemblages change, favoring species that grow rapidly in the presence of nutrients ("weed" species) over those which grow more slowly. There is some indication that eutrophication in salt waters is the source of the red tides that are a worldwide problem.

Under eutrophic conditions, the algae in the water significantly block the light passage, while in hypereutrophic conditions, the amount of biomass produced is so high that anoxic conditions occur, leading to fish kills. There are some indications that similar sorts of effects occur in terrestrial systems as well.

Table 4.6 Equivalency Factors for Acidifiers (Wenzel *et al* 1997).

Formula	**Conversion**	**M_w g.mol**	**n**	**EF kg SO_2/ kg substance**
SO_2	$SO_2 + H_2OH_2SO_3 \rightarrow 2H^+ + SO_3^{2-}$	64.06	2	1
SO_3	$SO_3 + H_2OH_2SO_4 \rightarrow 2H^+ + SO_4^{2-}$	80.06	2	0.80
NO_2	$NO_2 + 1/2H_2O + 1/4O_2 \rightarrow H^+ + NO_3H^-$	46.01	1	0.70
NO_x1	$NO_2 + 1/2H_2O + 1/4O_2 \rightarrow H^+ + NO_3H^-$	46.01	1	0 .70
NO	$NO + O_3 + 1/2H_2OH^+ + NO_3^- + 3/4O_2$	30.01	1	1.07
HCl	$HCL \rightarrow H^+ + Cl^-$	36.46	1	0.88
HNO_3	$HNO_3 \rightarrow H^+ + NO_3^-$	63.01	1	0.51
H_2SO_4	$H_2SO_4 \rightarrow 2H^+ + SO_4^{2+}$	98.07	2	0.65
H_3PO_4	$H_3PO_4 \rightarrow 3H^+ + PO_4^{3+}$	98.00	3	0.98
HF	$HF \rightarrow H^+ + F^-$	20.01	1	1.60
H_2S	$H_2S + 3/2O_2 + H_2O \rightarrow 2H^+ + SO_3^{2-}$	34.03	2	1.88
NH_3	$NH_3 + 2O_2 \rightarrow H^+ + NO_3^- + H_2O$	17.03	1	1.88

The ratio of carbon to nitrogen to phosphorus in aquatic biomass is 106:16:1 (Redfield 1942) , on an atomic basis. This ratio is the basis of combining nitrogen and phosphorus in calculating the eutrophication potential of emissions.

(Molar quantity of Nitrate + Nitrite + Ammonia)
× Redfield Ratio + Molar Quantity of Phosphate X = Eutrophication Indicator
(endpoint characterization factor (fresh, salt water))

Eutrophication is typically measured in PO_4 equivalents. The USEPA has set a concentration of 25 μg PO_4 L–1 as the level needed to protect fresh water aquatic ecosystems from eutrophication.

4.3.6 Energy

While inventory analysis involves the collection of data to quantify the relevant inputs and outputs of a product system, the accounting of electricity as a flow presents a unique challenge. The use of energy audits makes the idea of balancing energy flows around a process is a familiar one. However, in LCA the reporting of energy flows is in itself insufficient to perform a subsequent impact assessment. Ideally, the environmental impacts associated with energy generation should be captured in the approach. That is, the generation of electricity from fossil fuels should also show the contribution to the emission of global warming gases, solid waste (especially coal ash), etc. This type of detail also allows for the consideration of the use of waste materials in energy recovery operations. Also, the calculation of energy flow should take into account the different fuels and electricity sources used, the efficiency of conversion and distribution of energy flows as well as the inputs and outputs associated with generation and use of that energy flow. In addition, a more robust assessment may consider an evaluation of the specific sources of electrical power that are contributed to into the national energy grid on a more regional approach. This type of consideration is important in determining local impacts. For example, electricity that is produced in Maine is not used in California. Therefore, the impacts of electricity generation based on a national average may not be appropriate.

In the absence of a readily available model that can convert energy related inventory data into potential impacts based on the fuel source, a fallback position can be to look at the source of the total energy used and identifying what percentage is obtained from the national energy grid (which is mainly fossil fuels) and what percentage comes from other sources, such as the burning of waste materials. At this high level decision point, this information is appropriate and the approach fits the indicator-by-indicator comparison framework.

4.4 The Agri-Food Sector

The increasing activity in the agri-food sector, and the commensurate use of an ever growing quantity of freshwater and land in terms of pasture land, cropland, industrial space and relative infrastructures, deserves special attention. This increase in land and water use, as well as agro-chemical use, greatly influences the ecosystem. Initial approaches to the study of land use and water use in conjunction with LCA were developed over a decade ago (Lindeijer 2000; Lindeijer 2000; Lindeijer *et al* 2002; Müller-Wenk 1998; Owens 2002). Current methodologies dealing with these impact categories are still not fully representative of all the problems and aspects that can be encountered. As a consequence such methods need to be tailored and revised according to the agri-food product being considered during the LCA.

4.4.1 Land Use

The simplest way for LCAs of food products to evaluate the impact of land use is to consider only the area occupied (Schmidt 2008; Goedkoop *et al* 2009); however, in order to get a better picture of all environmental aspects, impacts on the quality of the soil (Saad *et al* 2011) or biotic potential and quality of the landscape (Mattsson *et al* 2000) should also be considered. The nature of these impact categories implies a difference in data type in terms of qualitative (e.g. biodiversity) versus quantitative (e.g. soil quality) data. There are therefore many kinds of land occupation/transformation indicators and no clear cut generally applicable impact methodology. Hence, it can be difficult to fully represent the sustainability of agri-food product systems.

"Midpoint Approach" indicators in general focus on soil quality indicators e.g. soil pH, phosphorous soil content etc. (Mattsson *et al* 2000) and are suited for comparisons between different land use activities. Soil Organic Matter (SOM) (Milá I Canals *et al* 2007) is considered a valid overall midpoint indicator but it gives no indication on other impacts such as soil erosion, compaction and salination. Müller-Wenk & Brandão (2010) considered the carbon transfer between vegetation/soil and land as a means of measuring the impact of land use.

"Endpoint Approach" indicators (Koellner and Scholz 2007; Koellner and Scholz 2008) take into account the naturalness of the system analyzed (e.g. biodiversity in terms of: potential disappeared fraction of vascular plant species) and are better related to other traditional impacts (e.g acidification and eutrophication). A recent approach to biodiversity assessment is found in (Koch *et al* 2010); however, it has only been tested on two indicators (grassland flora and grasshoppers) and needs to be broadened to include other indicators. In general "Endpoint Approach" indicators do not consider aspects such as the effects on human health or loss of unique (Mattsson *et al* 2000; Lindeijer 2000). New methodological approaches that take into consideration these impacts still need to be developed.

Since there are many possible land use indicators that can be considered, it is advisable to choose a set of indicators that best represents the agri-food environment being modeled. This will most likely force the LCA to follow a determined impact assessment path that might not represent all the possible impacts. On the other hand, the combination of all the indicators, considered in LCA study, into a final aggregate overall impact value is often difficult to achieve and at times unadvisable due to the very different nature of the collected information. For example, Mattsson when considering the land use for vegetable oil crops suggested that land use assessment is likely to be more descriptive and a step closer to an Environmental Impact Assessment study resulting in multiple impacts which should be associated with traditionally aggregated LCA results.

Site specificity is an issue that can cause great variability of the land use LCA results (Schryver *et al* 2010). Different regionalized datasets are already available, e.g. ReCiPe and Lime (Goedkoop *et al* 2009; Itsubo 2008) and should be used when possible in order to obtain realistic results. Recently Pfister *et al* (2010) used regionalized inventory and impact assessment data based on ecosystem vulnerability and net primary productivity of energy crops in order to demonstrate the tradeoffs between land and water use. This approach clearly demonstrates the variability of the regionalized results compared to the global averages; specifically although globally water only causes 25% of the global land impact, the land use impact is least in regions that have a high water use which, in this case, causes most of the damage. Work by Geyer *et al* 2010 and Nuñez *et al* 2010 developed methods that make use of geographical information systems (GIS) to couple site specific information with indicators regarding land use. Such bio-geographical differentiation approaches need to be further refined in order to be made mainstream but should nonetheless be implemented whenever possible in order to improve the quality of the agri-food LCA results.

Finally, performing impact assessment on the ecosystem relative to land use requires a measure of the impact referring to a baseline land where a previous state of the land is considered. The choice of the baseline can greatly influence the results of the LCA and therefore should be carefully chosen. For example, in Castanheira and Freire (2010) a choice of baseline land use of forest versus degraded grassland, when considering land use change to palm oil plantations, can lead from a reduction to a considerable increase in the impact on the ecosystem. Similarly in Peters *et al* (2010), when considering land use issues due to meat production, the author points out that land use transformation may or may not have originated due to the grazing activities causing baseline variability. Approaches used to define the baseline land can be found in Milá i Canals *et al* (2007).

4.4.2 Water Use

It is widely recognized that agricultural production is currently responsible for a large part of the global consumption of freshwater use. Even though freshwater use is a primary environmental concern, only recently there has been an increase in the study of aspects regarding the methodology for water use analysis in LCA (Milá i Canals *et al* 2008; Pfister *et al* 2009).

Initially in LCA freshwater use was only really considered at the inventory level by accounting for the amount of water withdrawn from the ground and surface without considering rain water since LCA originated as a tool for wet countries. A qualitative approach (water degradation), as opposed to quantitative (consumption), was then also introduced to report more meaningful inventory data in order to assess how much the utility of the returned water is impaired for either humans or ecosystems, as opposed to the effects of emissions to the aquatic environment assessed in conventional LCA (e.g. eutrophication and ecotoxicity). Recently, when measuring input and output data for all unit processes of the LCA, Boulay *et al* (2011) considered water categories by source, quality parameter and user as a means to quantify the elementary flows necessary for a subsequent evaluation the potential impacts of the degradative use of water in terms of loss of water functionality (Bayart *et al* 2010) for human users. Similarly Peters *et al* (2010) when performing hybrid LCA of Australian red meat production used a qualitative classification of water use.

Recent approaches to the impact assessment of water use can be found in Boulay *et al* 2011; Pfister *et al* 2009; Milá i Canals *et al* 2008. Since water footprint (WF) (Hoekstra and Chapagain 2008) and virtual water (VW) (Allan 1998) methodologies are applicable to products and are also similar to water use LCA methods at an inventory level, WF and VW are often used in conjunction with LCA to better evaluate the impacts of freshwater use. Approaches using modified LCA/WF/VW can be found Milà I Canals *et al* 2008, Pfister *et al* 2009, Ridoutt and Pfister 2010). In Jefferies *et al* (2010) tea and margarine production are considered using LCA and WF methodologies. The authors point out how the well-established databases and methodologies of LCA can help typical WF methodologies, whereas the concept of consumed water (as opposed to abstracted water) together with the methods of calculation of green and blue water in the absence of specific local data can improve the overall LCA estimations of the impacts of these agri-food products. Similarly Milà i Canals *et al* (2010) point out that WF accounting methods help to provide a richer picture of the total water consumption associated with growing of broccoli when assessing freshwater use impacts in LCA of such products in different parts of Europe. Some of the above mentioned authors when dealing with the impact of water use do not consider the green water component (as opposed to blue or grey water) and claim that it should be included in the land use impact assessment; others (Weidema & McGahan 2010) argue that it should be treated as a separate water resource. Such an aspect still seems to be an open issue.

As pointed out in Pfister *et al* (2010) water use and land use are closely correlated and hence are both affected by the site specificity of the assessment data. Pfister *et al* (2009) developed a method for assessing the environmental impacts, performed according to the framework of the Eco-indicator-99 assessment methodology, of freshwater consumption for cotton-textile production using regionalized characterization factors. Such an approach is also applied to agri-food water-intensive products such as vegetables and fruit (Pfister *et al* 2008) in different countries; this work clearly shows how results depend on irrigation requirements of the product. Most importantly the work indicates that more accurate specific regionalized results can contrast with average national impact values. Site specificity is also emphasiszed in the work by Hanafiah *et al* (2011). The derived characterization factors for water consumption and global warming based on freshwater fish species extinction can differ by orders of magnitude depending in the river being considered. Other recent approaches that consider regionalization aspects in water use can be found in Bayart *et al* (2010) and Boulay *et al* (2011).

In summary, work regarding water use is, in general, constantly evolving and has by no means reached a standard to be commonly used for all agri-food products. Nonetheless, the above cited approaches should be taken into consideration, whenever possible, by tailoring them to the specific agri-food LCAs that involve water use in order to improve the overall results.

4.4.3 Fertilizers and Pesticides

For the development of plants and to increase crops, agriculture makes extensive use of fertilizers and chemicals for weed and pest control. The quantity and quality of these products, in terms of their production and the resulting emissions, directly affect the results of LCI and LCIA. Fertilizer and pesticide use impacts strongly on impact categories such as global warming, acidification, eutrophication, human toxicity and ecotoxicity. For the food LCA practitioner, retrieval of data on the production of fertilizers and pesticides but also the estimation of the output of their use is particularly difficult. Many studies report that one of the most complex phase of a food LCA is just this.

In a review of wine LCA studies, Petti *et al* (2010) stated that one of the problems encountered by most of the analyzed studies was the difficulty in finding specific data and characterization factors for plant protection products and fertilizers used in the agricultural phase. As a consequence, an estimation of nutrients or pesticides releases through different dispersion models or assumptions was made. Margni *et al* (2002) stated that the assessment of the impact of pesticides on human health and ecosystem presents a certain degree of uncertainty as most of them, when used in agriculture, can be harmful for organisms which are not directly targeted, thus contaminating land and aquifers, and creating a risk for the population. Furthermore, the authors state that the impact of a pesticide depends on its interaction with the environment, its toxicity and quantity used.

To better understand the implications arising from the use of data on pesticides, the fate and behavior of pesticides in the environment must be known. The application of a herbicide or an insecticide may have different destinations: plants, air, soil, water and, indirectly, the terrestrial and aquatic wildlife and man. The environmental dispersion processes are surface runoff, leaching, volatilization, degradation and adsorption and desorption of pesticides in soil. The plants constitute the primary object of the treatments; the absorption of the pesticide by plants can be substantial. In applications of well-developed vegetation, for example, it can be intercepted and subsequently taken up to fifty percent of the amount of product used.

The air is simply a means of transport which the pesticide needs to reach the target. The phenomena concerning the passage of the pesticide through the air are the volatilization and drift. The volatilization of the pesticide is a transition to a vapor by sublimation and evaporation and depends largely on the nature of the compound and temperature. The drift is quite simply the physical transport of the pesticide or a part thereof at a point away from the application, mainly caused by the presence of wind during the distribution of the product and when treatments are carried out on the edge of the plot.

Most of the pesticide applied flows on the ground; pesticides in soil follow different paths depending on the complex interactions that are created between pesticide, soil, plants and weather conditions. The factors that most influence pesticide degradation into soil are represented by the physical and chemical properties of the product, the type and amount of microorganisms in the soil, moisture and soil temperature.

As with the soil, water can also be subject to receiving pesticide emissions. Pesticides can get to it through events such as meteoric run-off or leaching. In water bodies external to the agricultural land, contamination is mainly linked to the surface run-off. The percolation instead is due mainly to precipitation after application on the plants; the portion of the pesticide that is not absorbed by plants is in fact removed from their surface and ends in the ground. Pesticides move into the soil in various ways, among which the most important is the transportation in solution with water. The relative importance of each transport mechanism depends on the properties of pesticides, the amount of rainfall and chemical and physical characteristics of the soil. As a result of this flow a part of the pesticides can reach groundwater and finally the waters of rivers and wells. This phenomenon occurs when a fraction of the pesticide is removed and dissolved in water run-off and adsorbed on particles of eroded material. The magnitude of this fraction depends on the slope, on soil type, amount and intensity of rainfall.

From these considerations it follows that the fate of the pollutants associated with the use of a pesticide depends on site-specific data that are partly in contradiction with an analysis such as LCA that should be site independent. Moreover, the determination of the amount of pesticides remaining in the soil or their transformation products can be carried out with chemical and biological methods of analysis and with the implementation of predictive mathematical models, which have been developed especially in recent decades. In the agronomic and environmental chemistry literature there are, therefore, numerous dispersion models of pesticides in various environmental media and even software that perform the same type of analysis, developed in various countries. Some models are relative to the estimation of only a few effects, such as evaporation and degradation. These models start from some basic data concerning the characteristics of the pesticide, the characteristics of the soil, weather site and the characteristics of the plant to be treated. As quite evident, these data are site-specific and therefore the data obtained for the agricultural environment of a country and sold in commercial databases very often cannot be considered representative of other environments.

Whether dispersion models are employed or not, in any case many of the necessary data for the LCA, needed to follow the fate of pesticides, are difficult to obtain. In other cases the solution adopted for the emission estimate is derived from the analysis of the literature. Margni *et al* (2002) state that the fraction of the active ingredient entering the soil is assumed to be eighty five of the total applied quantity, assuming five per cent stays on the leaves in addition to ten per cent loss into the air while, few substances reach the groundwater and in most cases the pesticide run-off is less than ten per cent of the applied dose, based on literature (Audsley *et al* 1997). A more sophisticated model developed by Hauschild *et al* has been adapted to LCA (Hauschild, 2000; Birkved & Hauschild, 2006). It considers all types of dynamic behaviour during the emission of a pesticide, regarding emissions to air, groundwater and surface water in terms of elementary flows. Another approach to the problem is offered by Ecoinvent (2010) in which the entire amount of pesticide used as input shall be deemed emitted to the soil.

From the above analysis, it is clear that for the LCA implementer that is not an expert on pesticide chemistry or on mathematical dispersion models of pesticides, it is difficult to reach a scientific and reliable solution regarding the estimation of emissions from pesticides. We can conclude that there are two prevailing solutions: the first one concerns the use of dispersion models or literature data for the estimation of pesticide emissions to air, groundwater and surface water while the second one hypothesizes that the whole amount of pesticide used as an input is emitted to soil. These approaches naturally lead to different results both in terms of affected impact categories and absolute emission results; these results may further change, especially in comparative studies, if one uses different methods of impact assessment.

Furthermore, another problem regards the phase of impact assessment and, in particular, the existence of characterization factors of the pesticide used: whether a characterization factor for a given pesticide exists or not could enormously change the results of an LCA study.

Similar considerations can be made to estimate emissions from the use of fertilizers; in particular referring to the emissions of N_2O, NH_3, NO_3- to air, due to the use of nitrogen fertilizers, and of nitrate and phosphate leaching to the groundwater, due to the use of nitrogen and phosphorus fertilizers. The current extremely large variety of dispersion models and literature data (Brentrup *et al* 2000; ECETOC, 1994; Houghton, 1997) has created numerous approaches used by LCA scholars. Hence there is no current methodological standardization and it is therefore advisable to associate the LCA being carried out with a sensitivity or an uncertainty analysis in order to obtain a comprehensive picture of the overall results.

4.5 LCIA Models and Tools

Several LCIA models have been developed in recent years. While these models were designed to support conducting an ISO LCA, the methods within them adopted different modeling choices. These differences range from minor to significant. The types of modeling choices include the following:

- Modeling categories
- Midpoint and/or endpoint modeling
- Geographic focus
- Spatial differentiation
- Timeframe
- Assumptions about toxicity, etc.

Figure 4.5 shows how differently three LCIA methods present results on the same products.

The most often used LCIA models include the following:

- EPS 2000
- IMPACT World+ (updates IMPACT 2002+,LUCAS and EDIP)
- LIME: Life-cycle Impact assessment Method for Endpoint modeling
- ReCiPe (updates Eco-Indicator and CML)
- TRACI: The Tool for the Reduction and Assessment of Chemical and other environmental Impacts

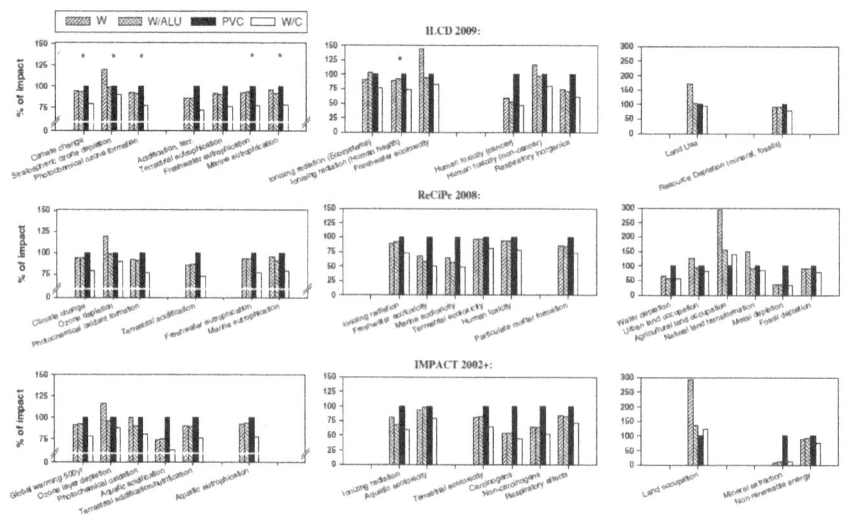

W: Pine wood W/ALU: Pine wood + aluminum PVC: Polyvinyl chloride with steel support W/C: Pine wood with glass fibre polyamide

Figure 4.5 Results of Modeling Four Window Types Using the ILCD Handbook, ReCiPe and Impact 2002+ (Owsianiak *et al* 2014).

Table 4.7 Common Life Cycle Impact Assessment Tools Sorted by Impact Category Indicator.

Distance-to-Target	**To Midpoint**	**To Damage or AoP**
Critical Volumina	CML (9+)	EPS (5)
Ecoscarcity (15)	EDIP (9)	Eco-indicator 99 (33)
	TRACI (12)	
	ILCD Handbook[(a)] (15)	ILCD Handbook[(a)] (3)
	Midpoint-Damage	
	IMPACT 2002+ (14-4)	
	LIME (11-4)	
	ReCiPe[(b)] (18-3)	
	IMPACT World+[(c)] (30-3)	

Numbers in parentheses (n) indicate the number of indicator categories.

[(a)] Midpoint and Damage impact categories are proposed, but not integrated in a consistent framework

[(b)] Created from a methodological update of CML 2001 and Eco-Indicator 99

[(c)] Created from a methodological update of IMPACT 2002+, LUCAS and EDIP

Table 4.7 summarizes these and other commonly used LCIA tools according to the impact indicator they aim to model. Although the ILCD Handbook is technically not a method, it is a collection of impact models, so it is included here.

Although studying the relevant documentation for an LCIA method is necessary in order to fully understand the models it contains, a qualitative comparison of key features and properties is helpful to for screening and making a first cut at choosing potentially relevant alternatives. Such a comparison is provided in Tables A.1 and A.2 in the chapter annex. This work was originally conducted by Hauschild *et al* (2013) in the context of the ILCD Handbook on LCIA (JRC 2011) and updated by Ralph Rosenbaum (2015) with further criteria and methods. These tables provide insights into the evolution of LCIA methods and, hence, the added value of updated methods intended to provide improved scientific and technical validity, etc.

References

Allan (1998) Virtual Water: A Strategic Resource Global Solutions to Regional Deficits. *Groundwater* 36(4):545–546.

Audsley A, Alber S, Clift R, Cowell S, Cretta P, Gaillard G, Hausheer J, Jolliet O, Kleijn R, Mortensen B, Pearce D, Roger E, Teulon H, Weidema B, & van Zeijts H (1997) *Harmonisation of Environmental Life Cycle Assessment for Agriculture, Final Report for Concerted Action AIR3-CT94-2028.* Silsoe, UK: Silsoe Research Institute.

Bare JC and Gloria TP (2006) Critical Analysis of the Mathematical Relationships and Comprehensiveness of Life Cycle Impact Assessment Approaches *Environ Sci Technol* 40(4): 1104-1113.

Bayart J-P, Bulle C, Deschênes L, Margni M, Pfister S, Vince F, and Koehler A (2010) A Framework for Assessing Off-Stream Freshwater Use in LCA *Int Life Cycle Assess* 15(5):439–453.

Birkved M and Hauschild M (2006) Pest LCI—A Model for Estimating Field Emissions of Pesticides in Agricultural LCA *Ecological Modelling* 198(3-4):433-451.

Boulay A-M, Bouchard C., Bulle C, Deschênes L, & Margni M (2011) Categorizing Water for LCA Inventory *Int J Life Cycle Assess* 16(7):639–651.

Boulay A-M, Bulle C, Bayart J, Deschenest L, and Margni M (2011) Regional Characterisation of Freshwater Use in LCA: Modelling Direct Impacts on Human Health *Environ Sci Technol* 45(20):8948–8957.

Brentrup F, Küsters J, Lammel J, and Kuhlmann H (2000) Methods to Estimate On-Field Nitrogen Emission from Crop Production as an Input to LCA Studies in the Agricultural Sector *Int. J. Life Cycle Assess* 5(6):349-357.

Castanheira É and Freire F (2010) CO2 and N2O Emissions from Palm Oil Plantations: Impacts of Land Use and Land Change. *In: Proceedings of the 7th International Conference on LCA in the Agri-Food Sector, Vol.1*, pp. 401-406. Bari, Italy.

ECETOC (1994) *Ammonia Emissions to Air in Western Europe-Technical Report no. 62.* Brussels, Belgium: European Chemical Industry, Ecology & Toxicology Center.

Ecoinvent (2010) Ecoinvent Data v2.2. Swiss Centre for Life Cycle Inventories.

Geyer R, Stoms, DM, Lindner JP, Davis FW, and Wittstock B (2010) Coupling GIS and LCA for Biodiversity Assessments of Land Use: Impact Assessment *Int J Life Cycle Assess* 14(7):692-703.

Geyer R, Stoms DM, Lindner JP, Davis FW, and Wittstock B (2010. Coupling GIS and LCA for Biodiversity Assessments of Land Use: Inventory Modeling. *Int J Life Cycle Assess* 15(5):454-467.

Goedkoop M, Heijungs R, Huijbregts M, Schryver AD, Struijs J, and Van Zelm R (2009) *ReCiPe 2008, A Life Cycle Impact Assessment Method which Comprises Harmonised Category Indicators at the Midpoint and the Endpoint Level- First Edition Report I: Characterisation.* http://www.lcia-recipe.net.

Hanafiah M, Xenopoulos M, Pfister S, Leuven R and Huijbregts M (2011) Characterization Factors for Water Consumption and Greenhouse Gas Emissions Based on Freshwater Fish Species Extinction *Environ Sci Technol* 45(12):5272–5278.

Hauschild M, Goedkoop M, Guinée J, Heijungs R, Huijbregts M, Jolliet O, Margni M, Schryver A, Humbert S, Laurent A, Sala S, Pant R (2013) Identifying Best Existing Practice for Characterization Modeling in Life Cycle Impact Assessment *Int J Life Cycle Assess* 18:683–697. doi: 10.1007/s11367-012-0489-5.

Hauschild M (2000) Estimating Pesticide Emission for LCA of Agricultural Products. In *Agricultural Data for Life Cycle Assessment* (pp. 64-79-Vol.2). Agricultural Economics Research Institute, The Hague: B Weidema, MJC Meeusen eds.

Hoekstra A and Chapagain A (2008) *Globalization of Water: Sharing the Planet's Freshwater Resources.* Oxford, UK: Blackwell Publishing.

Houghton J (1997) Greenhouse Gas Inventory Reporting Instructions, Volume 1-3, *Revised 1996 IPCC Guidelines for National Greenhouse Gas Inventories.* London, UK: IPCC - The Intergovernmental Panel on Climate Change.

Itsubo N (2008) *LIME Documentation. Eutrophication - Chapter 2.8 of 2. draft English translation of LIME documentation.*

Jefferies D, Muñoz I, King V, Canals LM., Hoekstra A, Aldaya M, and Ercin E (2010) A Comparison of Approaches to Assess Impact of Water Use in Consumer Products *In: Proceedings of the 7th International Conference on LCA in the Agri-Food Sector, Vol.1*, pp. 395-400. Bari, Italy.

Jolliet O, Müller-Wenk R *et al* (2004) The LCIA Midpoint-Damage Framework of the UNEP/SETAC Life Cycle Initiative *Int J Life Cycle Assess* 9(6):394-404.

JRC (2010) Framework and requirements for Life Cycle Impact Assessment (LCIA) Models and Indicators. In ILCD Handbook - International Reference Life Cycle Data System, European Commission - Joint Research Center.

JRC (2011) Recommendations for Life Cycle Impact Assessment in the European Context – Based on Existing Environmental Impact Assessment Models and Factors EUR 24571 EN; 159pp.

Koellner T and Scholz RW (2007) Assessment of Land Use Impacts on the Natural Environment. Part 1: An Analytical Framework for Pure Land Occupation and Land Use Change *Int J Life Cycle Assess* 12(1):16-23.

Koellner T and Scholz RW (2008) Assessment of Land Use Impacts on the Natural Environment. Part 2: Generic characterization factors for local species diversity in Central Europe *Int J Life Cycle Assess* 13(1):32-48.

SCS (2010) LCSEA Workshop on Draft National Standard for Evaluating Green Technologies, Climate Mitigation Strategies, Products and Services Using Life-Cycle Assessment, Washington DC, January 20, 2010.

Lindeijer E (2000) Biodiversity and Life Support Impacts of Land Use in LCA. *Journal of Cleaner Production, Vol.8*(Issue 4), 131-139.

Lindeijer E (2000) Review of Land Use Impact Methodologies *J Clean Prod* 8(4):273-281.

Lindeijer E, Müller-Wenk R and Stehen B (2002) Impact assessment of Resources and Land Use. In *Life-cycle Impact Assessment: Striving towards Best Practice* (pp. 11-64). Society of Environmental Toxicology and Chemistry, Pensacola, FL, USA: H.A. Udo de Haes, G. Finnveden, M. Goedkoop, M. Hauschild, E.G. Hertwich, P. Hofstetter, O. Jolliet, W. Klöpffer, W. Krewitt, E.W. Lindeijer, R. Müller-Wenk, S.I. Olsen, D.W. Pennington, J. Potting, and B. Steen eds.

Margni M., Rossier D., Crettaz P and Jolliet O (2002) Life Cycle Impact Assessment of Pesticides on Human Health and Ecosystems *Agriculture Ecosystems & Environment* 93:379-392.

Mattsson B, Cederberg C and Blix L (2000) Agricultural Land Use in Life Cycle Assessment (LCA): Case Studies of Three Vegetable Oil Crops *J Clean Prod* 8(4):283-292.

Milà i Canals L, Bauer C, Depestele J, Dubreuil A, Knuchel RF, Gaillard G, Michelsen O, Müller-Wenk R and Rydgren B (2007) Key Elements in a Framework for Land Use Impact Assessment in LCA *Int J Life Cycle Assess* 12(1): 5-15.

Milà i Canals L, Chenowith J, Chapagain A, Orr S, Antón A and Clift R (2008) Assessing Freshwater Use Impacts in LCA: Part 1: Inventory Modelling and Characterisation Factors for the Main Impact Pathways. *Int J Life Cycle Assess* 14(1):28–42.

Milà i Canals L, Chapagain A, Orr S, Chenoweth J, Antón A and Clift, R. (2010). Assessing Freshwater Use Impacts in LCA, Part 2: Case Study of Broccoli Production in the UK and Spain. *Int. J. Life Cycle Assess., Vol.15*(Issue 6), 598–607.

Milà i Canals L, Romanyà J and Cowell S (2007) Method for Assessing Impacts on Life Support Functions (LSF) Related to the Use of 'Fertile Land' in Life Cycle Assessment (LCA). *J Clean Prod* 15(15):426-1440.

Milà i Canals L, Sim S, Garcìa-Suàrez T, Neuer G, Herstein K, Kerr C, Rigarlsford G and King, H (2011) Estimating the Greenhouse Gas Footprint of Knorr *Int J Life Cycle Assess* 16(1):50-58.

Müller-Wenk R (1998) *Land Use—the Main Threat to Species. How to Include Land Use in LCA.* Switzerland: Universität of St.Gallen.

Müller-Wenk R and Brandão M (2010) Climatic Impact of Land Use in LCA - Carbon Transfers between Vegetation/Soil and Air. *Int J Life Cycle Assess* 15(2):172-182.

Nuñez M, Civit B, Muñoz P, Arena AP, Rieradevall J and Antón A (2010). Assessing Potential Desertification Environmental Impact in Life Cycle Assessment *Int J Life Cycle Assess* 15(1):67-78.

Owens JW (2002) Water Resources in Life-Cycle Impact Assessment *J Ind Ecol* 5(2):37-54.

Owsianiak M, Laurent A, Bjørn A, and Hauschild MZ (2014) IMPACT 2002+, ReCiPe 2008 and ILCD's Recommended Practice for Characterization Modelling in Life Cycle Impact Assessment: A Case Study-Based Comparison *Int J Life Cycle Assess* 19(5):1007–1021.

Peters G, Wiedemann S, Rowley H and Tucker R (2010) Accounting for Water Use in Australian Red Meat Production *Int J Life Cycle Assess* 15(3):311–320.

Petti L, Ardente F, Bosco S, De Camillis C, Masotti P, Pattara C, Raggi A and Tassielli G (2010) State of the Art of Life Cycle Assessment (LCA) in the Wine Industry. *In: Proceedings of the 7th International Conference on LCA in the Agri-Food Sector Vol 1*, pp. 493-498. Bari, Italy.

Pfister S, Curran M, Koehler A, and Hellweg S (2010) Trade-Offs between Land and Water Use: Regionalized Impacts of Energy Crops. *In: Proceedings of the 7th International Conference on LCA in the Agri-Food Sector* Vol.1, pp. 574-578. Bari, Italy.

Pfister S, Koehler A, & Hellweg S (2009) Assessing the Environmental Impacts of Freshwater Consumption in LCA *Environ Sci Technol* 43(11):4098–4104.

Pfister S, Stoessel F, Juraske R., Koehler A, & Hellweg S (2008) Regionalised LCIA of Vegetable and Fruit Production: Quantifying the Environmental Impacts of Freshwater Use. *In: Proceedings of the 6th int. Conference on LCA in the agrifood sector*, (pp. 16-21). Zurich, Switzerland.

Redfield AC (1942) The Processes Determining the Concentration of Oxygen, Phosphate, and other Organic Derivatives within the Depths of the Atlantic Ocean *Pap Phys Ocean Meteor* 9:22pp.

Ridoutt B and Pfister S (2010) A Revised Approach to Water Footprinting to make Transparent the Impacts of Consumption and Production on Global Freshwater Scarcity *Global Environmental Change* 20(1):113–120.

Rosenbaum R (2015) Chapter 2 Selection of Impact Categories, Category Indicators and Characterisation Models in Goal and Scope Definition, in the LCA Compendium, Goal and Scope Definition volume, Springer.

Saad R, Margni M, Koellner T, Wittstock B, & Deschênes L (2011) Assessment of Land Use Impacts on Soil Ecological Functions: Development of Spatially Differentiated Characterization Factors within a Canadian Context *Int J Life Cycle Assess* 16(3):1-14.

Schmidt J (2008) Development of LCIA Characterisation Factors for Land Use Impacts on Biodiversity *J Clean Prod* 16(18):1929-1942.

Schryver AD, Goedkoop M, Leuven R and Huijbregts M (2010) Uncertainties in the Application of the Species Area Relationship for Characterisation Factors of Land Occupation *Int J Life Cycle Assess* 15(7):682.

Schryver AD, Zelm R v, Goedkoop M and Huijbregts M (2010) Addressing Land Use and Ecotoxicological Impacts in Life Cycle Assessments of Food Production Technologies. In *Environmental assessment and management in the food industry: life cycle assessment and related approaches.* Cambridge: Woodhead Publishing: U. Sonesson, J. Berlin and Z. Friederike, edsWeidema & McGahan.

Wenzel H, Hauschild M and Alting L (1997) Environmental Assessment of Products, Vol. 1: Methodology, Tools and Case Studies in Product Development, Chapman & Hall, London.

Annex: Available Midpoint and Endpoint Characterization Methodologies

Table A.1 Detailed Midpoint Characterization Methodologies (updated by R. Rosenbaum) (JRC 2010, 2011)

LCIA methodology	CML 2002	TRACI 1.0	IMPACT 2002+	EDIP 2003	ReCiPe 2008	TRACI 2	ILCD/PEF/OEF	IMPACT World+
Reference	Guinée et al. (2002)	Bare et al. (2003)	Jolliet et al. (2003)	Hauschild and Potting (2003); Potting and Hauschild (2005)Hauschild and Wenzel (1998)	Goedkoop et al. (2012)	Bare (2011)	EC-JRC (2011)	Bulle et al. (2014)
Website	cml.leiden.edu/software/data-cmlia.html	epa.gov/nrmrl/std/traci/traci.html	impactmodeling.net		lcia-recipe.net		eplca.jrc.ec.europa.eu/?page_id=86	impactworldplus.org
Update of	CML 1992			EDIP 97	CML 2002	TRACI 1.0		IMPACT 2002+, EDIP 2003, and LUCAS
Uncertainties quantified								For each CF[1]
Normalisation factors	Netherlands (1997), Europe (1995), World (1995), production-based) (Huijbregts et al. 2003)	USA and USA+Canada (production-based) reference year 2005 (Lautier et al. 2010)	Western Europe (production-based)	Europe (production-based), reference year 2004 (Laurent et al. 2011)	Europe, global (production-based), reference year 2000 (Sleeswijk et al. 2008)	USA and USA+Canada (production-based) reference year 2008 (Ryberg et al. 2014)	European, global (production-based)	No normalization recommended at the mid-point level
Weighting scheme	Panel-based (~80 replies), deriving cultural perspectives: Individualist, Hierarchist, and Egalitarian based on Hofstetter (1998)			For emissions based on political targets for 2000, expressed as Political Target Person-Equivalents (PET), for resource consumption based on person reserves (PR) for known reserves in 1990	Panel-based (~80 replies), deriving cultural perspectives: Individualist, Hierarchist, and Egalitarian based on Hofstetter (1998)			No weighting recommended at the mid-point level

(*Continues*)

	LCIA methodology	CML 2002	TRACI 1.0	IMPACT 2002+	EDIP 2003	ReCiPe 2008	TRACI 2	ILCD/PEF/OEF	IMPACT World+
Climate change	**Characterisation model**	GWP[2] from IPCC[3] (2007a)	GWP[2] from IPCC[3] (2001)	GWP[2] from IPCC[3] (2001)	GWP[2] from IPCC[3] (2001)	GWP[2] from IPCC[3] (2007a)	GWP[2] from IPCC[3] (2007a)	GWP[2] from IPCC[3] (2007a)	GWP[2] from IPCC[3] 2013 (Myhre et al. 2013)
	Indicator	Radiative forcing	Radiative forcing	Radiative forcing	Radiative forcing	Radiative forcing	Radiative forcing	Radiative forcing	Radiative forcing
	Marginal/average	Average	Average	Average	Average	Average	Average	Average	Average
	Time horizon(s) [y]	100	100	500	100	100	100	100	≤100y, 100-500y
	Region modelled	Global	Global	Global	Global	Global	Global	Global	Global
	No. of substances	70	78	78	78	70	70	70	211
	Unit	CO_2 equivalents	CO_2 equivalents	CO_2 equivalents	CO_2 equivalents	CO_2 equivalents	CO_2 equivalents	CO_2 equivalents	CO_2 equivalents
Ozone depletion	**Characterisation model(s)**	ODP[4] from WMO[5] (2003)	ODP[4] from WMO[5] (1999)	ODP[4] from WMO[5] (2003)	ODP[4] from WMO[5] (2003)	ODP[4] from WMO[5] (2003)	ODP[4] from WMO[5] (2003)	ODP[4] from WMO[5] (2003)	ODP[4] from WMO[5] (2011)
	Indicator	Global degradation in stratospheric O_3 concentration	Global degradation in strato-spheric O_3 concentration	Global degradation in stratospheric O_3 concentration	Global degradation in stratospheric O_3 concentration	Global degrada-tion in strato-spheric O_3 concentration	Global degradation in strato-spheric O_3 concentration	Global degradation in strato-spheric O_3 concentration	Global degrada-tion in strato-spheric O_3 concentration
	Marginal/average	Average	Average	Average	Average	Average	Average	Average	Average
	Time horizon(s) [y]	Infinite	Infinite	Infinite	Infinite	Infinite	Infinite	Infinite	Infinite
	Region modelled	Global	Global	Global	Global	Global	Global	Global	Global
	No. of substances	20	20	20	20	20	20	20	23
	Unit	[6]CFC-11 equivalents	[6]CFC-11 equivalents	[6]CFC-11 equivalents	[6]CFC-11 equivalents	[6]CFC-11 equivalents	[6]CFC-11 equivalents	[6]CFC-11 equivalents	[6]CFC-11 equivalents

Respiratory inorganics	**Characterisation model**	PM[34] emissions to air considered in human toxicity impact category	De Hollander et al. (1999)	Hofstetter (1998)		van Zelm et al. (2008)	Same as TRACI 1.0	[7]Humbert (2009)	Humbert et al. (2011) (for fate and exposure) and Gronlund et al. (in prep.) (for effects)
	Fate/exposure modelling	PM[34] emissions to air considered in human toxicity impact category	CALPUFF model, mechanistic, closed, multimedia LCA model	Empirical data		Mechanistic, option for low, undefined and high stack emission	Same as TRACI 1.0	[7]Humbert (2009) (USEtox (Rosenbaum et al. 2008) is used for primary PM and CO, Greco et al. (2007) for secondary PM from SO_2 and NOx, Van Zelm et al. (2008) for secondary PM from NH_3 and RiskPoll (Rabl and Spadaro 2004) to differentiate among high-stack, low-stack, and ground-level emissions of primary PM for urban and rural conditions, respectively)	Mechanistic (same as in ILCD/PEF/OEF but with updated parameters)

(*Continues*)

LCIA methodology	CML 2002	TRACI 1.0	IMPACT 2002+	EDIP 2003	ReCiPe 2008	TRACI 2	ILCD/PEF/OEF	IMPACT World+
Effect modelling		endpoint-based indicator, dose-response (chronic mortality, acute mortality, acute respiratory morbidity, acute cardiovascular morbidity). 10.9 DALY per case mortality is used	Dose-response (chronic mortality, acute mortality, acute respiratory morbidity, acute cardiovascular morbidity)		Endpoint-based indicator, dose-response (chronic mortality, acute mortality, acute respiratory morbidity, acute cardiovascular morbidity); chronic bronchitis not considered		Epidemiological studies (dose-response (chronic mortality, acute mortality, acute respiratory morbidity, acute cardiovascular morbidity)	Epidemiological studies (dose-response (chronic mortality, acute mortality, acute respiratory morbidity, acute cardiovascular morbidity)
Effect modelling		endpoint-based indicator, dose-response (chronic mortality, acute mortality, acute respiratory morbidity, acute cardiovascular morbidity). 10.9 DALY per case mortality is used	Dose-response (chronic mortality, acute mortality, acute respiratory morbidity, acute cardiovascular morbidity)		Endpoint-based indicator, dose-response (chronic mortality, acute mortality, acute respiratory morbidity, acute cardiovascular morbidity); chronic bronchitis not considered		Epidemiological studies (dose-response (chronic mortality, acute mortality, acute respiratory morbidity, acute cardiovascular morbidity)	Epidemiological studies (dose-response (chronic mortality, acute mortality, acute respiratory morbidity, acute cardiovascular morbidity)
Marginal/average		Marginal and Average give same results	Marginal and Average give same results		Marginal and Average give same results		Marginal and Average give same results	Marginal and Average give same results

	Emission compartment(s)		Air (point source (proxy for high stack) and mobile source (proxy for low stack)	Air		Air (rural, urban, and undefined)		Air (high stack, low stack, ground-level, undefined; remote, rura, urban and undefined environnement)	Air (high stack, low stack, ground-level, undefined; remote, rura, urban and undefined envi-ronnement)
	Time horizon, discounting		Infinite	Infinite		No timeframe, no discounting		No timeframe, 3% discounting for future cost	No timeframe, no discounting
	Region modelled		USA	Europe		Europe		Generic	Generic and continental
	Level of spatial differentiation					Urban and rural archetypes		High stack, low stack, ground-level, undefined; remote, rura, urban and undefined environnement	High stack, low stack, ground-level, unde-fined; remote, rura, urban and undefined environnement
	No. of substances		7 (TSP[8], PM10, PM2.5, NO, NO_2 NO_x/nitrate and SO_2/sulfate)[9]	6 (primary PM10 and PM 2.5, SO_2, NO_2, secondary PM10 from NO_2, SO_2)[9]		4 (primary PM10; secondary PM10 from NH_3, NO_x, SO_2)[9]		5 (primary PM2.5, primary PM10, secondary PM from NO_x and SO_2, and CO)[9]	5 (primary PM2.5, pri-mary PM10, secondary PM from NO_x and SO_2, and CO)[9]
	Unit		PM10 equivalents[9]	PM2.5 equivalents[9]		PM10 equivalents[9]	PM2.5 equivalents[9]	PM2.5 equivalents[9]	PM2.5 equivalents[9]

(*Continues*)

LCIA methodology		CML 2002	TRACI 1.0	IMPACT 2002+	EDIP 2003	ReCiPe 2008	TRACI 2	ILCD/PEF/OEF	IMPACT World+
Photochemical ozone formation/respiratory organics	Characterisation model	Derwent et al. (1998), Scenario reflecting realistic worst case for British Isles	Norris (2003)	POCP[10] from Jenkin et al. (1999)	Hauschild et al. (2006)	van Loon et al. (2007), Vautard et al. (2007), vanZelm et al. (2008)	Based on TRACI 1.0 but updated	Recommended: LOTOS-EUROS fate model as used in ReCiPe, resulting indicator: tropospheric ozone concentration increase (no effects)	van Loon et al. (2007), Vautard et al. (2007), vanZelm et al. (2008)
	Fate modelling	POCP[10], detailed fate modelling of individual VOCs[11]	Change in Maximum Incremental Reactivity (MIR) (Carter 2000), ASTRAP model for source-receptor matrix, factor for NO_x based on US average estimate	POCP[10] from Jenkin et al. (1999)	Regression model derived from RAINS	LOTOS-EUROS model 2007	Updated MIR (Maximum Incremental Reactivity) based on Carter (Bare 2011)		LOTOS-EUROS model 2007
	Exposure modelling			Population densities and average daily inhalation	Increased exposure of humans or vegetation above critical threshold, population density	Population densities within grid cells, atmospheric concentration of O_3 and average daily inhalation	Similar to TRACI 1.0		Population densities within grid cells, atmospheric concentration of O_3 and average daily inhalation
	Effect modelling		Human health only, linear dose-response model		Exceedance for exposure above critical level for humans and vegetation based on WHO[12] guidelines (AOT60 for human health, AOT40 for vegetation)	Human health only, linearity assumed with no threshold (based on WHO[12] recommendation)			Human health only, linearity assumed with no threshold (based on WHO[12] recommendation)
	Marginal/average	Marginal (ΔO_3 per marginal ΔVOC[11])	Marginal for fate, average for effect	Marginal and Average give same results	Marginal increase in long term ozone levels	Marginal (linearity checked up to 10% increase)			Marginal (linearity checked up to 10% increase)

	Emission compartment(s)	Air	Air	Air	Air	Air			Air
	Region modelled/ valid	North Western Europe	USA	Europe	Europe	Europe			Europe
	Level of spatial differentiation		US states		European countries	Global generic			Global generic
	No. of substances	127 (VOCs[11], CH_4, SO_2, NO, NO_2 and CO)	~580	~130	4 (nmVOC[13], CH_4, No_x and CO)	2 (nmVOC[13] and NO_x)	~1200		134
	Unit	C_2H_4 equivalents	NO_x equivalents	ethylene equivalents	person*ppm*h and m^2*ppm*h	kg nmVOC[13] equivalents	NO_x equivalents		kg nmVOC[13] equivalents
Ionizing radiation	**Characterisation model**	Frischknecht et al. (2000)		Same as CML 2002, but without effect model, indicator: human exposure level		Same as CML 2002, but without effect/ damage model, indicator: human exposure level		Recommended: Frischknecht et al. (2000) for human health but excluding damage assessment (only effect) Interim use of method by Garnier-Laplace et al. (2008; 2009) for ecosystem impacts possible	Same as CML 2002, but without effect model, indicator: human exposure level Garnier-Laplace et al. (2008; 2009) for ecosystem impacts

(*Continues*)

LCIA methodology	CML 2002	TRACI 1.0	IMPACT 2002+	EDIP 2003	ReCiPe 2008	TRACI 2	ILCD/PEF/OEF	IMPACT World+
Fate modelling	Dreicer et al. (1995), using routine atmospheric and liquid discharges into rivers in French nuclear fuel cycle including surrounding conditions. For globally dispersed radionuclides simplified models are used.							
Exposure modelling	Effective dose [Sv][14] via inhalation, ingestion, external irradiation, based on human body equivalence factors for α-, β-, γ-radiation, and neutrons							
Effect modelling	Dose-response functions directly based on human subjects exposed in Nagasaki and Hiroshima (extrapolated to low-dose exposure)							
Time horizon(s) [y]	100, 100000							
Emission compartment(s)	Air, water							
Region modelled/ valid	Global, Europe (fate based on French conditions)							

	Level of spatial differentiation								
	No. of substances	31 (21 radionuclides to outdoor air, 13 to water, 15 to ocean)							26 for human health and 13 for ecosystem quality impacts
	Unit	DALY		C-14 Bq equivalents[15]		man.Sv/kBq [14 15]			Bq C-14 equivalents[15]
Human toxicity	**Diseases considered**	Cancer/non-cancer	Cancer/non-cancer	Cancer/non-cancer	Cancer/non-cancer	Cancer/non-cancer	Cancer/non-cancer	Cancer/non-cancer	Cancer/non-cancer
Human toxicity	**Characterisation model**	USES-LCA 1.0 (Huijbregts et al. 2000)	CalTOX 4.0 (Hertwich et al. 2001; McKone et al. 2001)	IMPACT 2002 (Pennington et al. 2005)	EDIP (Hauschild and Potting 2003)	USES-LCA 2.0 (van Zelm et al. 2009)	USEtox (Rosenbaum et al. 2008)	USEtox (Rosenbaum et al. 2008)	USEtox (Rosenbaum et al. 2008)
Human toxicity	**Fate modelling**	Mechanistic, nested, multimedia, mass-balance model (developed for LCA)	Mechanistic, closed, multimedia, mass-balance model (developed for ERA)	Mechanistic, nested, multimedia, mass-balance model (developed for LCA)	Key property, partial fate	Mechanistic, nested, multimedia, mass-balance model (developed for LCA)	Mechanistic, nested, multimedia, mass-balance model (developed for LCA)	Mechanistic, nested, multimedia, mass-balance model (developed for LCA)	Mechanistic, nested, multimedia, mass-balance model (developed for LCA)
Human toxicity	**Exposure modelling**	Inhalation, various ingestion pathways	Inhalation, various ingestion pathways, dermal uptake	Inhalation, various ingestion pathways	Inhalation, various ingestion pathways	Inhalation, various ingestion pathways	Inhalation, various ingestion pathways	Inhalation, various ingestion pathways	Inhalation including indoor exposure (Hellweg et al. 2009; Wenger et al. 2012), various ingestion pathways including pesticide residues in food from Fantke et al. (2011a; 2011b; 2013)
Human toxicity	**Effect modelling**	[16]ED50	linear slope factor based on RfD[17]	[18]ED10	[18]ED10	[16]ED50	[16]ED50	[16]ED50	[16]ED50

(Continues)

LCIA methodology	CML 2002	TRACI 1.0	IMPACT 2002+	EDIP 2003	ReCiPe 2008	TRACI 2	ILCD/PEF/OEF	IMPACT World+
Marginal/average	Marginal (non-linear effect factor)	Marginal and Average give same results	Marginal and Average give same results	Marginal and Average give same results	Marginal (non-linear effect factor)	Marginal and Average give same results	Marginal and Average give same results	Marginal and Average give same results
Time horizon	Infinite, 100 years for metals	Infinite	Infinite	Infinite	Infinite, 100 years for metals or all substances in Individualist perspective	Infinite	Infinite	Infinite, ≤100 y, >100 y for metals
Time horizon	Infinite, 100 years for metals	Infinite	Infinite	Infinite	Infinite, 100 years for metals or all substances in Individualist perspective	Infinite	Infinite	Infinite, ≤100 y, >100 y for metals
Emission compartment(s)	Rural air, urban air, freshwater, agricultural soil, natural soil, industrial soil	Air, freshwater, soil	Air, freshwater, marine water, soil	Air, freshwater, soil	Rural air, urban air, freshwater, agricultural soil, natural soil, industrial soil	Rural air, urban air, water, agricultural soil, natural soil	Rural air, urban air, water, agricultural soil, natural soil	Rural air, urban air, water, agricultural soil, natural soil
Region modelled	Europe, version available for various continents	USA	Europe	Europe	Europe	Generic	Generic	Generic global average + 9 parameterised sub-continents
Level of spatial differentiation								sub-continental level (Kounina et al. 2014)
No. of substances	~860	~380	~800	~180	~1000	~1250	~1250	~1250
Unit	1,4-DCB equivalents[19]	2,4-D equivalents[20]	chloroethylene equivalents	m^3 (volume of poisoned compartment)	1,4-DCB equivalents[19]	cases or CTU_h[21]	cases or CTU_h[21]	cases or CTU_h[21]

Ecotoxicity	**Ecosystems considered**	Freshwater, freshwater sediment, marine, marine sediment, terrestrial	Freshwater, terrestrial	Freshwater, marine, terrestrial	Freshwater, terrestrial	Freshwater, marine, terrestrial	Freshwater	Freshwater	Freshwater, interim factors for marine and terrestrial
	Characterisation model	USES-LCA 1.0 (Huijbregts et al. 2000)	CalTOX 4.0 (Hertwich et al. 2001; McKone et al. 2001; Bare et al. 2003)	IMPACT 2002 (Pennington et al. 2005)	EDIP1997, combined with site dependent factors (Tørsløv et al. 2005)	USES-LCA 2.0 (van Zelm et al. 2009)	USEtox (Rosenbaum et al. 2008)	USEtox (Rosenbaum et al. 2008)	USEtox (Rosenbaum et al. 2008)
	Fate/exposure modelling	Mechanistic, nested, multimedia, mass-balance model (developed for LCA)	Mechanistic, closed, multimedia, mass-balance model (developed for ERA)	Mechanistic, nested, multimedia, mass-balance model (developed for LCA)	Key property, partial fate	Mechanistic, nested, multimedia, mass-balance model (developed for LCA)	Mechanistic, nested, multimedia, mass-balance model (developed for LCA)	Mechanistic, nested, multimedia, mass-balance model (developed for LCA)	Mechanistic, nested, multimedia, mass-balance model (developed for LCA)
	Effect modelling	Most sensitive species	Most sensitive species	Average toxicity	Most sensitive species	Average toxicity	Average toxicity	Average toxicity	Average toxicity
	Marginal/average	Marginal and Average give same results	Marginal and Average give same results	Marginal and Average give same results	Marginal and Average give same results	Marginal (non-linear effect factor)	Marginal and Average give same results	Marginal and Average give same results	Marginal and Average give same results
	Time horizon	Infinite	Infinite	Infinite	Infinite	Infinite, 100 years for metals	Infinite	Infinite	Infinite, ≤100 y and >100 y for metals
	Emission compartment(s)	Air, freshwater, marine water, agricultural soil, industrial soil	Air, freshwater	Air, freshwater, soil	Air, freshwater, soil	Air, freshwater, marine water, agricultural soil, natural soil	Air, freshwater, marine water, agricultural soil, natural soil	Air, freshwater, marine water, agricultural soil, natural soil	Air, freshwater, marine water, agricultural soil, natural soil
	Region modelled	Europe	USA	Europe	Generic	Europe	Generic	Generic	Generic global average + 9 parameterised sub-continents
	Level of spatial differentiation								sub-continental level (Kounina et al. 2014)
	No. of substances	~170	~160	~430	~190	~2650	~2550	~2550	~2550
	Unit	1,4-DCB equivalents[19]	2,4-D equivalents[20]	triethylene-glycol equivalents	m^3 (volume of poisoned compartment)	1,4-DCB equivalents[19]	PAF[22] in $[m^3*day]$ or CTU_e[23]	PAF[22] in $[m^3*day]$ or CTU_e[23]	PAF[22] in $[m^3*day]$ or CTU_e[23]

(*Continues*)

LCIA methodology		CML 2002	TRACI 1.0	IMPACT 2002+	EDIP 2003	ReCiPe 2008	TRACI 2	ILCD/PEF/OEF	IMPACT World+
Eutrophication	Ecosystems considered	Freshwater, terrestrial	Freshwater	Freshwater	Freshwater, terrestrial	Freshwater, marine	Same as TRACI 1.0 but with additional substances covered	Freshwater, marine, terrestrial	Freshwater, marine
	Fate modelling	No mechanistic fate model, mineralisation with full release of bioavailable nutrients, Redfield ratio assumed for N/P ratio, no advection, no distinction between sensitive and insensitive recipients	Mechanistic (ASTRAP for atmospheric transport, WSAG topological network for water transport) except for NH_3 (NO_x matrices have been used as placeholder), no advection, no N or P removal, distinction between sensitive and insensitive recipients	Same as CML 2002, but distinguishing P-limited watershed, N-limited watershed, and undefined watershed (modeled as a 50% P-limited and a 50% N-limited watershed)	Atmospheric fate and transport (RAINS model pre 2000 with detailed transport and fate model (applying EMEP)), no advection, removal of nutrients in hydrological cycle (using CARMEN model as fixed removal ratio for N- and P-compounds in different emission scenarios)	Mechanistic fate model EUTREND for atmospheric and CARMEN for waterborne emissions (Struijs et al. 2009), no advection leaving Europe, fixed removal ratio for N and P in different emission scenarios		Recommendation for freshwater and marine: EUTREND model (Struijs et al. 2009) as implemented in ReCiPe [kg P to freshwater and kg N to freshwater for freshwater and marine respectively]	Freshwater: River global network model from Vörösmarty et al. (2000a; 2000b) + fate model accounting for advection, retention and water use at the 0.5°x0.5° resolution scale Marine : GEOSchem model (from NASA http://map.nasa.gov/GEOS_CHEM.html) for atmospheric fate and transport until coastal zones at the 2°x2.5° resolution scale

Exposure modelling		Distinction of N- and P- limited recipients		No distinction between freshwaters and marine waters, no critical levels in water, exceedance of critical load in soil considered	Distinction between exposure of N- and P-limited systems		Recommendation for terrestrial: Accumulated Exceedance based on Seppälä et al. (2006), Posch et al. (2008), not included in any other LCIA method Distinction between exposure of N- and P-limited systems	Distinction between exposure of N- and P-limited systems
Effect modelling					Linear dose-response relationship according to limiting nutrient			Freshwater = P limited Generic effect factor from Struijs et al. (2011) Marine = N limited Generic effect factor from Jolliet (unpublished)
Marginal/average	Marginal	Average		Marginal	Marginal		Marginal	Marginal
Emission compartment(s)	Air, freshwater, marine water, soil	Air, freshwater		Air, freshwater	Air, freshwater, soil		Air, freshwater, marine water	Air, freshwater, marine water
Time horizon	Infinite	Infinite		Infinite	Infinite		Infinite	Infinite
Region modelled/ valid	Global generic	USA		Europe	Europe		Europe	Global

(*Continues*)

LCIA methodology	CML 2002	TRACI 1.0	IMPACT 2002+	EDIP 2003	ReCiPe 2008	TRACI 2	ILCD/PEF/ OEF	IMPACT World+
Level of spatial differentiation		US states		European countries	Spatial differentiation not found important for aquatic eutrophication (factor 3-7 between European countries)		Generic, countries for NH_3 and NO_2	CFs[26] available at global, continental, country, and fine scale (0.5°x0.5° resolution scale for freshwater eutrophication 2°x2.5° for marine eutrophication)
No. of substances	13 (8 N-compounds, 4 P compounds, COD[24])	11 (to air: NO, NO_2, NO_x, NH_3 to water: NH_4^+, N, NO_3^-, PO_4^{3-}, P, COD[24])		Freshw.: 12 (5 to air, 7 to water; 9 N-comp, 3 P-comp); Terrestrial: SO_2, SO_3,H_2SO_4, H_2S, NO, NO_2, NO_x, HNO, NH_3, HCL, HF	4 (N-total, P-total, NO_x and NH_3, but differentiation of N-total and P-total according to source)		Freshwater: P, H_3PO_4, P-total; marine: NH_3, NH_4^+, NO_3^-, NO_2^-, NO_2, NO, N-total; terrestrial: NH_3, NH_4^+, NO_2, NO, NO_3^- (to air)	8 for freshwater eutrophication, 16 for marine eutrophication
Unit	PO_4^{3-} equivalents	PO_4^{3-} equivalents		Freshwater: NO_3^- equivalents; terrestrial: m^2 unprotected ecosystem	kg P to freshwater, kg N to freshwater (for marine eutrophication)		kg P to freshwater, kg N to freshwater (for marine eutrophication), mol N-equivalents for terrestrial	kg PO_4 P-lim equivalents for freshwater eutrophication, kg N N-lim equivalents for marine eutrophication

Acidification	**Ecosystems considered**	Terrestrial	Terrestrial		Terrestrial	Terrestrial (forest soil)		Terrestrial	Terrestrial, freshwater
	Characterisation model	Huijbregts et al.(2001)	Norris (2003)		Potting et al. (1998)	van Zelm et al. (2007) EUTREND, SMART 2		Seppälä et al. (2006) and Posch et al. (2008)	Roy et al. (2012a; 2012b; 2014)
	Fate modelling	Mechanistic atmospheric fate model (RAINS model dated before 2000), linear increase in sensitive area change according to emission scenario	Atmospheric fate and deposition on land (ASTRAP model dated before 2000), no soil sensitivity to acidifying deposition	Same as CML 2002	Mechanistic atmospheric fate model, linear increase in sensitive area change, sensitive areas considered, incl. with limited buffer capacity, discounting for deposition in area above critical load	Mechanistic atmospheric fate model (EUTREND model) including deposition on land (SMART 2 model)	Same as TRACI 1.0 but with additional substances covered	Mechanistic atmospheric fate model including deposition on land (EMEP model)	GEOSchem model (from the NASA) for atmospheric fate and transport at the 2°x2.5° resolution scale for both terrestrial and freshwater acidification + receiving environment fate model for freshwater acidification
	Exposure modelling	Sensitive areas considered, incl. with limited buffer capacity, acidification potential modelled with slope inversly proportional to critical load and applied above and below it				Sensitive areas consider magnitude of deposition above critical load and areas with limited buffer capacity for forests (extrapolated to other ecosystems)		linear increase (sensitive area change), sensitive areas, incl. with limited buffer capacity, acidification potential modelled as the exceedence above the critical load, potency	Terrestrial acidification: Soil sensitivity factor giving the change in soil solution H^+ concentration due to a change in the atmospheric deposits of pollutant using the PROFILE geochemical steady-state model

(*Continues*)

LCIA methodology	CML 2002	TRACI 1.0	IMPACT 2002+	EDIP 2003	ReCiPe 2008	TRACI 2	ILCD/PEF/OEF	IMPACT World+
Effect modelling	Dose-response slope based on a midpoint hazard risk ratio, similar to PEC / PNEC, and applied above and below critical load (validity still need to be verified as some doubts exist on the relevance of the dose-response curve, as the slope over buffer capacity depends on the buffer capacity itself							
Marginal/average	Marginal	Average		Marginal	Marginal		Marginal	Marginal
Emission compartments	Air, freshwater, soil	Air		Air	Air		Air	Air
Time horizon(s) [y]	1990 and 2010 emission scenario	present		1990 and 2010 emission scenario	20, 50, 100 and 500		2002 and 2010 emission scenario	2010 emission scenario
Region modelled	Europe	North America		Europe	Europe + country specific validity		Europe	Global
Level of spatial differentiation	European countries	US counties		European countries			European countries	CFs[26] available at global, continental, country, and fine scale (2° x 2.5° resolution scale)
No. of substances	4 (NH_3, NO_x (NO_2 and NO), SO_2)	8 (H_2S, SO_x, NO, NO_2, NO_x, HCL, HF, NH_3)		11 (SO_2, SO_3, H_2SO_4, H_2S, NO, NO_2, NO_x, HNO, NH_3, HCL, HF)	4 (NO_x (NO, NO_x), NH_3, SO_2)		4 (SO_2, NO_x (NO, NO_2), NH_3)	15
Unit	kg SO_2 equivalents	H^+ equivalents		kg SO_2 equivalents	BS[25] $[m^2 * y]$		mol H^+ equivalents	kg SO_2 equivalents

Land use								
Aspects considered			Based on Eco-Indicator 99 endpoints: obtained by dividing damage factors from Eco-Indicator 99 by damage factor of organic arable land from Eco-Indicator 99		Land surface used		Soil quality/function	Biodiversity, Ecosystem services: erosion resistance capacity/potential (ERP), fresh water recharge potential (FWRP), mechanical filtration potential (MWFP), chemical filtration potential (CWFP), carbon sequestration potential (CSP), biotic production potential (BPP)
Characterisation model					None, hence no distinction8 of different species composition between land use types		Mila i Canals (2007), based on soil organic matter (SOM), influencing soil functions like fertility, pH buffering, soil structure, water retention capacity	Biodiversity: de Baan et al. (2013a; 2013b) Ecosystem services: updated models from Saad et al. (2011; 2013), Müller-Wenk & Brandão (2010) and Brandão & Milà i Canals (2013)

(*Continues*)

LCIA methodology	CML 2002	TRACI 1.0	IMPACT 2002+	EDIP 2003	ReCiPe 2008	TRACI 2	ILCD/PEF/OEF	IMPACT World+
Marginal/average					Average		Not described	Not described
Region modelled					Global generic		No direct CFs[26] developed	Global
Level of spatial differentiation							No direct CFs[26] developed	16 WWF biomes for biodiversity; 36 Holdridge lifezones for ERP, MWPP, PCWPP, FWRP; 16 & 122 WWF biomes & ecozones respectively for CSP; 12 IPCC climate zones for BPP
No. of land use types					3 (2 occupation (competition in agricultural area, competition in urban area) and 1 transformation)		No specific CF[26] developed, example list with input data is provided, more figures can be found in literature	8 for biodiversity; 36 for ERP, MWPP, PCWPP, FWRP; 3 for CSP; 8/26 for BPP
Unit			m^2 equivalents of organic arable land * y		m^2 occupation or transformation		Soil organic matter (SOM)	Biodiversity in ha equivalents of arable land * y, ERP in ton/ha/y, FWRP in mm/y, MWFP in cm/d, CWFP in $cmol/kg_{soil}$, CSP in tCO_2, BPP in tC/ha/y

Resources	**Aspects considered**	Abiotic: amount of non-renewable resources (fossil fuels and miner-als) extracted	Same as Eco-Indicator 99 endpoint for fossil use only	Based on Eco-Indicator 99 endpoints: obtained by dividing damage factors from Eco-Indicator 99 by damage factor of iron in ore from Eco-Indicator 99	Abiotic: amount of non-renewable resources (fossil fuels and minerals) extracted Biotic: wood extraction	Abiotic: non-renewable resources (fossil fuels and miner-als), based on change in availability of high grade minerals and fossil resources, mining of bulk res. covered by land use, water only as inventory parameter Biotic: use of agricultural, silvicul-tural biotic resources covered by land use	Same as TRACI 1.0	Recommendation for minerals and fossil: CML 2002	Fossil fuels and mineral resources, approaches based on dissipation of resource functional-ity (instead of depletion of stocks) assuming that extraction does not contribute to functional-ity loss and therefore only dissipation of a resource has an impact
	Characterisation model	Guinée and Heijungs (1995)based on extraction rates and using ultimate reserves, average rock is used			Hauschild and Wenzel (1998): amount of resource extracted is compared to 1990 extraction levels in person equivalents (frac-tion of resource which can be exploited econom-ically is used (can be much smaller than the ultimate reserves))	Kirkham and Rafer (2003)			Fatemi et al. (unpub-lished), de Bruille et al. (unpublished)

(*Continues*)

	LCIA methodology	CML 2002	TRACI 1.0	IMPACT 2002+	EDIP 2003	ReCiPe 2008	TRACI 2	ILCD/PEF/OEF	IMPACT World+
	Time horizion	Infinite			For renewables annual regeneration used to determine supply horizon	Minerals: infinite Fossil fuels: before and after 2030			Infinite
	Region modelled	Global			Global	Global			Global
	No. of resource types	82 elements			36	20 (minerals) + 34 (fossil fuels)			n/a
	Unit	Abiotic Depletion Potential (ADP) [dimensionless]		kg equivalents of iron in ore	Person reserve - quantity of resource available per person (according to economically exploitable reserve)	Marginal increase of extraction costs			MJ deprived, kg deprived
Water use	**Aspects considered**							Water scarcity, no impacts	Water deprivation
	Characterisation model(s)							Water consumption as in Swiss Ecoscarcity (Frischknecht et al. 2009), amount of non-renewable resources extracted using 6 levels of water scarcity in a region (comparing extraction with water input)	Based on Boulay et al. (2011), representing water stress, i.e. level of competition among users (anthropogenic and ecosystems) due to physical stress of the resource, addressing quality, seasonal variations, and distinguishing between surface and groundwater
	Marginal/average							Not described	Not described

	Region modelled							Several countries	Global
	Level of spatial differentiation							Water scarcity given for many important countries	Country, watershed
	Unit							m^3 equivalent	m^3 deprived

[1] Characterisation factor
[2] Global Warming Potential
[3] Intergovernmental Panel on Climate Change
[4] Ozone Depletion Potential
[5] World Meteorological Organisation
[6] Chlorofluorocarbon
[7] The information given in ILCD handbook and related documents is incorrect. The correct models are given in Humbert (2009) – the ILCD recommended approach – and are reflected here as well.
[8] Total Suspended Particulate matter
[9] PM - Particulate Matter (with diameters up to 2.5μm and 10μm respectively)
[10] Photochemical Ozone Creation Potential
[11] Volatile Organic Compounds
[12] World Health Organisation
[13] Non-Methane Volatile Organic Compounds
[14] Sievert, unit of ionizing radiation dose
[15] Becquerel, unit of radioactivity (1 Bq = 1 disintegration per second)
[16] Effective Dose affecting 50% of tested individuals
[17] Reference Dose (US-EPA's acceptable daily oral exposure to the human population likely to be without risk of deleterious effects during a lifetime)
[18] Effective Dose affecting 10% of tested individuals
[19] Dichlorobenzene
[20] Dichlorophenoxyacetic acid
[21] Comparative Toxic Unit for humans
[22] Potentially Affected Fraction of species (not an actual unit but a fraction of 1)
[23] Comparative Toxic Unit for ecosystems
[24] Chemical Oxygen Demand
[25] Base Saturation

Table A.2 Detailed Endpoint Characterization Methodologies (updated by R. Rosenbaum) (JRC 2010, 2011)

LCIA methodology	EPS 2000	Eco-Indicator99	IMPACT 2002+	LIME 1.0 (2003)	LIME 2.0 (2008)[1]	ReCiPe 2008	ILCD/PEF/OEF	IMPACT World+
Reference	Steen (1999)	Goedkoop and Spriensma (2000)	Jolliet et al. (2003)	Itsubo and Inaba (2003)	Itsubo and Inaba (2012)	Goedkoop et al. (2012)	EC-JRC (2011)	Bulle et al. (2014)
Website	cpmdatabase.cpm.chalmers.se/AboutDatabase_2.htm	pre-sustainability.com/eco-indicator-99-manuals	impactmodeling.net	aist-riss.jp/main/modules/groups_alca/content0005.html?ml_lang=en	lca-forum.org/english	lcia-recipe.net	eplca.jrc.ec.europa.eu/?page_id=86	impactworldplus.org
Update of					LIME 1.0	Eco-Indicator99		IMPACT 2002+
Uncertainties quantified					full uncertainty information based on Monte Carlo Analysis			For each CF[2]
Normalisation factors	Integrated in CFs[2]	Western Europe (production-based)	Western Europe (production-based)	Integrated in CFs[2]	Integrated in CFs[2]	Europe, global (production-based)	European, global (production-based)	Global, planetary boundaries
Weighting scheme	Integrated in CFs[2]	Panel-based cultural perspectives: Individualist, Hierarchist, and Egalitarian		Willingness to pay [Yen] (Conjoint analysis of interview survey of ~400 Japanese people (Kanto region) on opinions about environmental policy)	Same as LIME 1.0, but for ~1000 people from all over Japan and with improved statistical analysis	Cultural perspectives: Individualist, Hierarchist, and Egalitarian		Stepwise 2006 monetization values for DALYs[55] and PDF[56] in [m^2.y]
Areas of protection	Human health [YOLL][3], biodiversity [NEX][4], abiotic resources [kg], ecological productivity [kg]	Human health [DALY][5], ecosystem quality [PDF in km^2*y][6], resources [MJ surplus energy]	Human health [DALY][5], ecosystem quality [PDF in m^2*y][6], resources [MJ surplus energy], climate change [CO_2 eq]	Human health [DALY][5], social welfare [Yen], biodiversity [EINES][7], primary production [kg-DW][8]	Human health [DALY][5], social assets [Yen], biodiversity [EINES], primary production [kg-DW][8]	Human health [DALY][5], Ecosystems [species*y], resources [surplus cost]		Human health [DALY][5], Ecosystems [PDF in m^2*y][6], ecosystem and resources servicec [$]

	LCIA methodology	EPS 2000	Eco-Indicator99	IMPACT 2002+	LIME 1.0 (2003)	LIME 2.0 (2008)[26]	ReCiPe 2008	ILCD/PEF/OEF	IMPACT World+
Climate change	Characterisation model(s)	(IPCC 1990)	GWP[9] from IPCC[10] (1995)		GWP[9] from IPPC[10] (2001)	GWP[9] from IPPC[10] (2007b)	GWP[9] from IPCC[10] (2007b), de Schryver et al. (2009)		GTP[11] from IPCC[10] 2013 (Myhre et al. 2013), de Schryver et al. (2009) with corrected values for "semi natural areas"
	Human health effects	Thermal stress, flooding, malaria, starvation (no diarhea)	Thermal stress, vector borne diseases, flooding (no starvation)		Heat and cold stress, vector borne diseases, malnutrition, natural disasters	Heat and cold stress, malaria, dengue, disaster damage, malnutrition, hunger	Thermal stress, flooding, malaria, starvation, diarhea		Thermal stress, flooding, malaria, starvation, diarhea
	Ecosystem effects			only midpoint CF			Species loss based on global assessment of ecosystem studies in different regions and different species groups	No recommendation, but interim use of ReCiPe 2008 model	Species loss based on global assessment of ecosystem studies in different regions and different species groups
	Biotic resources effects	Decrease in crop production, increase wood production and species loss			Links to crop production and fish production	Agricultural production, energy consumption, land disappearance			Ecosystem service loss (interim)
	Marginal/average	Average	Marginal		Average	Average	Marginal		Marginal
	Time horizon(s) [y]	100	200		100	100	100		100, [100-500]
	Region modelled	Global	Global		Global	Global	Global		Global
	Level of spatial differentiation								
	No. of substances	106	38		n/a	n/a	70		211
	Unit	Willingness to pay (WTP)	DALY[5]		DALY[5] and Yen (crop loss)	DALY[5] and Yen	DALY[5] and PDF[6] in [m²*y]		DALY[5], PDF[6] in [m²*y] and $

(*Continues*)

Ozone depletion	**Characterisation model(s)**	ODP[12] from WMO[13] (1999)	ODP[12] from WMO[13] (1999) and data from TOMS satelite	ODP[12] from WMO[13] (2003) and US-EPA	ODP[12] from WMO[13] (1999)	Same as LIME 1.0	ODP[12] from WMO[13] (2003)	No recommendation, but interim use of ReCiPe 2008 model	ODP[12] from WMO[13] (2011)
	Human health effects	Skin cancer	Skin cancer and cataract	Same as Eco-Indicator 99 (Egalitarian scenario), but with updated ODP	Cataract and skin cancer (malignant melanoma, basal cell carcinoma, squamous cell carcinoma)		Skin cancer and cataract (latter not for individualist perspect.) based on the AMOUR 2.0 model (van Dijk et al. 2008; Den Outer et al. 2008)		Skin cancer and cataract based on the AMOUR 2.0 model (van Dijk et al. 2008; Den Outer et al. 2008)
	Biotic resources effects				Net primary productivity for coniferous forests, agriculture (soybean,rice, greenpea, mustard) and phytoplancton at high latitudes				
	Marginal/average	Average	Marginal		Marginal		Marginal		Average
	Time horizon(s) [y]	Infinite	500		Infinite		Infinite		Integrated from 2007 to 2100
	Region modelled	Global	Global		Global		Global		Global
	Level of spatial differentiation								
	No. of substances	20	25	95	19		22		23
	Unit	DALY[5] converted to monetary value using willingness to pay	DALY[55]	DALY[55]	DALY[5], NPP[14] and kg Wood		DALY[5]		DALY[5]

Respiratory inorganics	**Characterisation model**		Hofstetter (1998)	Same as Eco-Indicator 99	Hofstetter (1998), Ikeda (2001)	Hofstetter (1998), Ikeda (2001), Pope et al. (2002)	van Zelm et al. (2008)	Recommendation: DALY[5] based on Humbert (2009)	Humbert et al. (2011) (for fate and exposure) and Gronlund et al. (unpublished) (for effects)
	Fate/exposure modelling		Mechanistic		Mechanistic	Mechanistic atmospheric fate (plume model and puff model) distinguishing chimneys and vehicles source types	Mechanistic, distinguishing low, undefined and high stack emission		Mechanistic (same as in ILCD/PEF/OEF but with updated parameters)
	Effect modelling		Dose-response (chronic mortality, acute mortality, acute respiratory morbidity, acute cardiovascular morbidity, etc.)		Dose-response (chronic mortality, acute mortality, acute respiratory morbidity, acute cardiovascular morbidity, etc.)	Similar to LIME 1.0 but disease and death rates were increased for some substances/endpoints according to newer data from Pope et al. (2002)	Dose-response (chronic mortality, acute mortality, acute respiratory morbidity, acute cardiovascular morbidity) ; chronic bronchitis not considered		Dose-response, mass-based, surface-based, and number-based
	Marginal/average		Marginal and Average give same results		Marginal and Average give same results	Marginal and Average give same results	Marginal and Average give same results		Marginal and Average give same results
	Emission compartment		Air		Air	Air (from chimney, from automobile)	Air (for low, undefined and high stack emission)		Air (high stack, low stack, ground-level, undefined; remote, rura, urban and undefined environnement)

(*Continues*)

LCIA methodology	EPS 2000	Eco-Indicator99	IMPACT 2002+	LIME 1.0 (2003)	LIME 2.0 (2008)[26]	ReCiPe 2008	ILCD/PEF/OEF	IMPACT World+
Time horizon, discounting		no timeframe, no discounting		Infinite	Same as LIME 1.0	no timeframe, no discounting		no timeframe, no discounting
Region modelled		Generic continent		Average atmospheric conditions for 7 regions of Japan		Europe		Generic and continental
Level of spatial differentiation				7 Japanese regions				high stack, low stack, ground-level, undefined; remote, rura, urban and undefined environnement
No. of substances		4 (primary PM10; secondary PM10 from NH_3, NO_x, SO_2; no values for PM2.5)[15]		6 (primary PM10 and PM 2.5, SO_2, NO_2, secondary PM10 from NO_2, SO_2)[15]		4 (primary PM10; secondary PM10 from NH_3, NO_x, SO_2)[15]		5 (primary PM2.5, primary PM10, secondary PM from NO_x and SO_2, and CO)[15]
Unit		DALY[5]		DALY[5]		DALY[5]		DALY[5]
Human health effects	Morbidity and severe morbidity	Acute mortality	Same as Eco-Indicator 99 (Egalitarian scenario)	Acute mortality, morbidity, no consideration of NO_x	Same as LIME 1.0	Only acute mortality, chronic effects disregarded due to lack of empirical evidence	Recommendation: ReCiPe 2008 model	Constant damage factor for acute mortality, chronic effects disregarded due to lack of empirical evidence
Ecosystem effects				All relevant mechanisms, no consideration of NOx				

	Characterisation model	POCP[16] (Lindfors et al. 1994)	Hofstetter (1998)		Ozone conversion equivalency factors (OCEF), Schere & Demerjian, (1984), corrected by Uno and Wakamatsu (1992)		van Loon et al. (2007), Vautard et al. (2007), vanZelm et al. (2008)		van Loon et al. (2007), Vautard et al. (2007), vanZelm et al. (2008)
Photochemical ozone formation/respiratory organics	**Fate modelling**	Based on old Swedish version of POCP[16] (Lindfors et al. 1994)	POCP[16] from Jenkin et al. (1999)		Average atmospheric conditions for 7 regions of Japan		LOTOS-EUROS model 2007		LOTOS-EUROS model 2007
	Exposure modelling		Population densities and average daily inhalation		Population distribution for human health, integrated in dose-response model for other AOPs[17]		Population densities within grid cells, atmospheric concentration of O_3 and average daily inhalation		Population densities within grid cells, atmospheric concentration of O_3 and average daily inhalation
	Effect modelling	Based on global average situation	Linear exposure-response functions		Linear exposure-response functions, also considering crops and natural vegetation, wood		Linearity assumed, no threshold and only including acute effects		Human health only, linearity assumed with no threshold (based on WHO[18] recommendation)
	Marginal/average	Average	Marginal and Average give same results		Marginal (ΔO_3 per marginal ΔVOC[19] for 8 VOC[19] archetypes)[19]		Marginal (linearity checked up to 10% increase)		Marginal (linearity checked up to 10% increase)
	Emission compartment	Air	Air		Air		Air		Air
	Region modelled/valid	Global damage data, European fate data (POCP[16])	Europe		Japan		Europe		Europe
	Level of spatial differentiation								

(*Continues*)

LCIA methodology		EPS 2000	Eco-Indicator99	IMPACT 2002+	LIME 1.0 (2003)	LIME 2.0 (2008)[26]	ReCiPe 2008	ILCD/PEF/OEF	IMPACT World+
	No. of substances	67 (nmVOCs[20] and No_x)	130		8 archetypes from which individual VOCs[19] can be estimated		2 (nmVOC[20] and NO_x) but factors for individual VOCs[19] can be determined applying POCPs[16], e.g. from CML 2002		134
	Unit	DALY[5] converted to monetary value using willingness to pay	DALY[5]		DALY[5]		DALY[5]		DALY[5]
Ionizing radiation	Characterisation model		Frischknecht et al. (2000)	Same as Eco-Indicator 99 (Egalitarian scenario)			Same as Eco-Indicator 99	No recommendation, but interim use of Frischknecht et al. (2000) as used in Eco-Indicator 99, IMPACT 2002+, ReCiPe	Same as CML 2002, but without effect model, indicator: human exposure level Garnier-Laplace et al. (2008; 2009) for ecosystem impacts
	Fate modelling		Dreicer et al. (1995), using routine atmospheric and liquid discharges into rivers in French nuclear fuel cycle including surrounding conditions. For globally dispersed radionuclides simplified models are used.						
	Exposure modelling		Effective dose [Sv][21] via inhalation, ingestion, external irradiation, based on human body equivalence factors for α-, β-, γ-radiation, and neutrons						

	Effect modelling		Dose-response functions directly based on human subjects exposed in Nagasaki and Hiroshima (extrapolated to low-dose exposure)						
	Time horizon(s) [y]		100, 100000						
	Emission compartment		Air, water						
	Region modelled/ valid		Global, Europe (fate based on French conditions)						
	No. of substances		31 (21 radionuclides to outdoor air, 13 to water, 15 to ocean)						26 for human health and 13 for ecosystem quality impacts
	Unit		DALY[5]						DALY[5], PDF[6] in [m^2*y]

(*Continues*)

LCIA methodology		EPS 2000	Eco-Indicator99	IMPACT 2002+	LIME 1.0 (2003)	LIME 2.0 (2008)[26]	ReCiPe 2008	ILCD/PEF/OEF	IMPACT World+
Human toxicity	**Diseases considered**	Cancer/non-cancer	Cancer	Cancer/non-cancer	Cancer, oral chronic non-cancer diseases	Cancer, oral and inhalatory chronic non-cancer diseases, sick house syndrome via indoor exposure to formaldehyde, NO_x, SO_2, PM	Cancer/non-cancer	Recommendation for cancer effects: USEtox model (Rosenbaum et al. 2008) combined with DALYs[5] from Huijbregts et al. (2005) No recommendation for non-cancer effects, but interim use of USEtox model (Rosenbaum et al. 2008) combined with DALYs[5] from Huijbregts et al. (2005)	Cancer/ non-cancer
	Characterisation model	None	EUSES 1.0 (EC and RIVM 1996)	IMPACT 2002 (Pennington et al. 2005)	Modified version of IMPACT 2002 model	Modified version of IMPACT 2002 model	USES-LCA 2.0 (van Zelm et al. 2009)		USEtox (Rosenbaum et al. 2008)
	Fate modelling	None	Mechanistic, mass-balance model (developed for risk assessment)	Mechanistic, mass-balance model (developed for LCA)			Mechanistic, mass-balance model (developed for LCA)		Mechanistic, mass-balance model (developed for LCA)
	Exposure modelling	some pathways missing	inhalation, various ingestion pathways	inhalation, various ingestion pathways			inhalation, various ingestion pathways		Inhalation including indoor exposure (Hellweg et al. 2009; Wenger et al. 2012), various ingestion pathways including pesticide residues from Fantke et al. (2011a; 2011b; 2013)

	Effect modelling	RfD[22], safety factors	Unit-Risk concept	ED10[23]	Unit-risk concept for cancer ED10[23] for oral chronic diseases	Improved models and data, including inhalation exposure to heavy metals based on epidemiological data, sick house syndrome effect modelling via indoor inhalation volume-response curve based on clinic survey	ED50[24]		ED50[24]
	Marginal/average	Average, Sweden, Europe or USA taken as reference	Marginal and Average give same results	Marginal and Average give same results			Marginal (non-linear effect factor)		Marginal and Average give same results
	Time horizon	n/a	Infinite	Infinite	Similar to IMPACT 2002+	Similar to IMPACT 2002+	Infinite, 100 years for metals or all substances in Individualist perspective		Infinite, ≤100 y, >100 y for metals
	Emission compartment	n/a	Air, water, soil	Air, water, soil			Rural air, urban air, water, agricultural soil, natural soil		Rural air, urban air, water, agricultural soil, natural soil
	Region modelled	Global average, calculated based on Swedish data	Europe	Europe	Japan	Japan	Europe		Global average + 9 sub-continents
	Level of spatial differentiation								sub-continental level (Kounina et al. 2014)
	No. of substances	~57	~10	~800	~800	~1000	~1000		~1250
	Unit	DALY[5] converted to monetary value	DALY[5]	DALY[5]	DALY[5]	DALY[5]	DALY[5]		DALY[5]

(Continues)

	LCIA methodology	EPS 2000	Eco-Indicator99	IMPACT 2002+	LIME 1.0 (2003)	LIME 2.0 (2008)[26]	ReCiPe 2008	ILCD/PEF/OEF	IMPACT World+
Ecotoxicity	**Ecosystems considered**	n/a	Freshwater, terrestrial	Freshwater, marine, terrestrial	Freshwater, terrestrial	Freshwater, terrestrial	Freshwater, marine, terrestrial	No recommendation	Freshwater, marine, terrestrial
	Characterisation model	None (red list species pot. threatened by chemicals)	EUSES 1.0 (EC and RIVM 1996)	IMPACT 2002 (Pennington et al. 2005)	Modified version of IMPACT 2002 model	Modified version of IMPACT 2002 model	USES-LCA 2.0 (van Zelm et al. 2009)		USEtox (Rosenbaum et al. 2008)
	Fate/exposure modelling	None	Mechanistic, mass-balance model (developed for risk assessment)	Mechanistic, mass-balance model (developed for LCA)			Mechanistic, mass-balance model (developed for LCA)		Mechanistic, mass-balance model (developed for LCA)
	Effect modelling	None	Most sensitive species	Average toxicity			Average toxicity		Average toxicity
	Marginal/average	Marginal and Average give same results	Marginal and Average give same results	Marginal and Average give same results			Marginal (non-linear effect factor)		Marginal and Average give same results
	Time horizon	n/a	Infinite	Infinite			Infinite, 100 years for metals		Infinite, $\leq$100 y and >100 y for metals
	Emission compartment	n/a	Air, water, soil	Air, water, soil			Air, water, agricultural soil, natural soil		Air, water, agricultural soil, natural soil
	Region modelled	Global average	Europe	Europe	Japan	Japan	Europe		Global average + 9 sub-continents
	Level of spatial differentiation								sub-continental level (Kounina et al. 2014)
	No. of substances	~45	~45	~430	n/a	n/a	~2650		~2550
	Unit	NEX[14]	PDF[16] in [m^2*y]	PDF[16] in [m^2*y]	EINES[25]	EINES[25]	PDF[6] in [y/t]		PDF[6] in [m^2*y]

Acidification	**Ecosystems considered**		Terrestrial (incl. wetlands and swamps)	Same as Eco-Indicator 99 (Egalitarian scenario)	Terrestrial	Terrestrial	Terrestrial	No recommendation, but interim use of ReCiPe 2008 model	Terrestrial, freshwater, marine
	Characterisation model		SMART, MOVE		Hayashi et al. (2004)	Hayashi et al. (2004) and others	van Zelm et al. (2007) EUTREND, SMART 2		Roy et al. (2012a; 2012b; 2014)
	Fate modelling		No atmospheric fate model, assuming 60% of emissions deposited on European natural soil (no advection leaving Europe)		Atmospheric fate model (mixing mechanistic and empirical approaches - Euler-type model for SO_2 and No_x, empirical values for rest of chemicals with emission and deposition data between 1991-1997) including deposition on land, domestic average	Similar to LIME 1.0 but dividing Japan into 6 zones, including geographical fluctuation, calculating source attribution by zone based on the ratio of terrestrial area and marine area	Mechanistic atmospheric fate model (EUTREND model) including deposition on land (SMART 2 model)		GEOSchem model (from NASA http://map.nasa.gov/GEOS_CHEM.html) for atmospheric fate and transport at the 2°x2.5° resolution scale for both terrestrial and freshwater acidification + receiving environment fate model for freshwater acidification

(*Continues*)

LCIA methodology	EPS 2000	Eco-Indicator99	IMPACT 2002+	LIME 1.0 (2003)	LIME 2.0 (2008)[26]	ReCiPe 2008	ILCD/PEF/OEF	IMPACT World+
Exposure modelling		Change in soil pH for a marginal deposition increase based on Dutch conditions		Sensitive areas consider magnitude of deposition above critical load and areas with limited buffer capacity for representative soils in Japan	Naturally produced SO_2 excluded from emissions used for calculation of source attribution of SO_2	Sensitive areas consider magnitude of deposition above critical load and areas with limited buffer capacity for forests (extrapolated to other ecosystems)		Terrestrial: soil sensitivity factor giving the change in soil solution H^+ concentration due to a change in the atmospheric deposits of pollutant using the PROFILE geochemical steady-state model (Roy et al. 2012a)
Effect modelling		Linear dose-response relationship, endpoint as impacts on higher plants in the Netherlands		Change in NPP of forestry plant species per change H+ deposition: [NPP] / [H^+_dep], linear function between H+ deposition and NPP, latter calculated as a function of soil acidification and Al_3^+ concentration	Similar to LIME 1.0 but updated modelling and added biodiversity of terrestrial (plant) ecosystems	Change in Potentially Not Occurring Fraction of forestry plant species per change in Base Saturation: [PNOF] / [BS], linear function for BS >0.15 based on 240 vascular plants species		Terrestrial: site specific effect factor based on the biome regression models from Azevedo et al. (2013) Freshwater: location-specific effect factor linking change in H^+ concentration in a lake to change in potentially disappeared fraction (Roy et al. 2014)

									Marine: dose response curve determined for pH effect on different calcifying organisms using available litterature data allowing to assess their EC50[27] (de Schryver et al., unpublished)
	Marginal/average		Marginal		Marginal	Same as LIME 1.0	Marginal		Marginal
	Emission compartment		Air		Air		Air		Air
	Time horizon(s) [y]		Infinite		Infinite		20, 50, 100 and 500		Infinite
	Region modelled		Europe + country specific validity/ Dutch validity for midpoint-endpoint modelling		Japan		Europe + country specific validity/ Dutch validity for midpoint-endpoint modelling		Global
	Level of spatial differentiation								CFs[2] available at global, continental, country, and fine scale (2° x 2.5° resolution scale)
	No. of substances		7 (SO_2, SO_3, SO_x, NH_3, NO, NO_2, No_x)		5 (SO_2, NO, NO_2, HCL, NH_3)	Same as LIME 1.0	4 (No_x (NO, NO_x), NH_3, SO_2)		15
	Unit		PDF[6] in [m^2*y]		NPP[28]	NPP[14] and Yen	PDF[16] in [m^2*y]	species*y	PDF[16] in [m^2*y]

(*Continues*)

LCIA methodology		EPS 2000	Eco-Indicator99	IMPACT 2002+	LIME 1.0 (2003)	LIME 2.0 (2008)[26]	ReCiPe 2008	ILCD/PEF/OEF	IMPACT World+
Land use	**Aspects considered**	biodiversity (only red list species) and primary wood productivity of forest	Biodiversity	Same as Eco-Indicator 99 (Egalitarian scenario)	Biodiversity and net primary production of an area	Similar to LIME 1.0 but with updated data covering 30 types of vegetation	Biodiversity	No recommendation, but interim use of ReCiPe 2008 model	Biodiversity, Ecosystem service loss: erosion resistance capacity/potential (ERP), fresh water recharge potential (FWRP), mechanical filtration potential (MWFP), chemical filtration potential (CWFP), carbon sequestration potential (CSP), biotic production potential (BPP)
	Characterisation model	(Jarvinen and Miettinen 1987)	Koellner (2001), Species richness (based on vascular plants), species-area relationship		NPP based on Chikugo model. Biodiversity based on red species list and life expectancy (study by Matsuda et al. (2003))		Koellner (2001), Species richness (based on vascular plants), species-area relationship		Biodiversity: de Baan et al. (2013a; 2013b) Ecosystem services: updated models from Saad et al. (2011; 2013), Müller-Wenk & Brandão (2010) and Brandão & Milà i Canals (2013)

Marginal/average	Not described	Marginal		Marginal		Marginal		Not described
Region modelled	Global, but based only on Swedish data	Mid Europe, based on Swiss data		Japan, for NPP[64], overseas areas also considered		North West-Europe, based on British and Swiss data		Global
Level of spatial differentiation								16 WWF biomes for biodiversity; 36 Holdridge lifezones for ERP, MWPP, PCWPP, FWRP; 16 & 122 WWF biomes & ecozones respectively for CSP; 12 IPCC climate zones for BPP
No. of land use types	3 (arable land, forest for forest, and forest for roads)	16 land occupations + 11 land conversions		n/a	9	18 land occupations (including 3 intensities for arable areas) + 4 land conversions		8 for biodiversity; 36 for ERP, MWPP, PCWPP, FWRP; 3 for CSP; 8/26 for BPP
Unit	For biodiversity: NEX[4]; for primary production: 1kg dry wood	PDF[6] in $[m^2*y]$		EINES[7] and NPP[14]	EINES[7] and NPP[14]	PDF[6] in $[m^2*y]$		Biodiversity affecting Ecosystem quality endpoint in PDF[6] in $[m^2*y]$, Ecosystem services (FWRP, MWFP, CWFP, BPP) affecting all three andpoints given in DALY[5], PDF[6] in $[m^2*y]$, \$, Ecosystem service ERP affecting Ecosystem and resources services in \$

(Continues)

	LCIA methodology	EPS 2000	Eco-Indicator99	IMPACT 2002+	LIME 1.0 (2003)	LIME 2.0 (2008)[51]	ReCiPe 2008	ILCD/PEF/OEF	IMPACT World+
Resources	**Aspects considered**	Abiotic: non-renewable resources (fossil fuels and minerals) based on resource in average rock, water extraction (irrigation, drinking) Biotic: wood extraction, fish and meat extraction	Abiotic: non-renewable resources (fossil fuels and minerals), based on (long term) change in availability of high grade minerals and fossil resources, mining of bulk resources covered by land use Biotic: use of agricultural, silvicultural biotic resources covered by land use	Abiotic: non-renewable resources: minerals as modelled in Eco-Indicator 99 and fossil fuels modelled in a specific way (no stock size included), mining of bulk resources covered by land use Biotic: use of agricultural, silvicultural biotic resources covered by land use	Abiotic: minerals, fossil resources, fossil fuels Biotic: living resources (terrestrial ecosystem)	Same as LIME 1.0 but with improved models and data	Abiotic: non-renewable resources (fossil fuels and minerals), based on change in availability of high grade minerals and fossil resources, mining of bulk resources covered by land use, water only as inventory parameter Biotic: use of agricultural, silvicultural biotic resources covered by land use		Endpoint: ecosystem and resources services affected due to functional deprivation of fossil fuels and mineral resources, approaches based on dissipation of resource functionality (instead of depletion of stocks) assuming that extraction does not contribute to functionality loss and therefore only dissipation of a resource has an impact
	Characterisation model	Numerous sources from 1990-2000, not all input data is traceable	(de Vries 1988) (Müller-Wenk 1998)	For minerals see Eco-Indicator 99; for fossils: Ecoinvent database as of 2003	n/a		Kirkham and Rafel (2003)		Fatemi et al. (unpublished), de Bruille et al. (unpublished)
	Time horizon	Long time frame, weighting/ normalization based on future technologies	5x the historical extraction before 1990	Minerals: 5x the historical extraction before 1990. Fossil fuels: infinite	n/a		Minerals: infinite Fossil fuels: before and after 2030		n/a
	Region modelled	Global	Global	Global	n/a		Global		Global

	No. of resource types	67 (minerals) + 3 (fossils) + 3 (biotic) + 1 (water)	12 (minerals) + 9 (fossil fuels)	22 (minerals) + 9 (fossil fuels)	n/a		20 (minerals) + 34 (fossil fuels) + 5 (water use)		n/a
	Unit	kg of element or resource used	Surplus energy [MJ] needed to extract one kg extra element or one MJ fuel.	Surplus Energy and primary energy [MJ]	Yen, EINES[57], and NPP[64]		Marginal increase of extraction costs		$
Water use	**Aspects considered**								Human health, ecosystem quality, ecosystem and resources services
	Characterisation model(s)								Boulay et al. (2011), van Zelm et al. (2010) Hanafiah et al. (2011)
	Ecosystem effects								Terrestrial: based on van Zelm et al. (2010) regionalised estimates using the likelihood of shallow vs. deep groundwater extraction per country Freshwater: based on Hanafiah et al. (2011) as global average value
	Human health effects								Boulay et al. (2011)

(*Continues*)

	LCIA methodology	EPS 2000	Eco-Indicator99	IMPACT 2002+	LIME 1.0 (2003)	LIME 2.0 (2008)[51]	ReCiPe 2008	ILCD/PEF/OEF	IMPACT World+
	Region modelled								Global
	Level of spatial differentiation								CFs[52] available at global, continental, country, and for human health also at fine scale (watershed resolution scale)
	Unit								DALY[55], PDF[56] in $[m^2*y]$, $
Rare or emerging impact categories	**Noise**					From transport applying fate (increase in noise levels due to additional cars/km - distinguishing day/night and small/large cars), exposure (distribution of exposed population with back-ground levels), effect (volume-response curve based on social survey), and damage due to sleep disorder and conversa-tion disorder			

	Waste					User cost, biodiversity of terrestrial ecosystems			
	Thermal water pollution								Based on Verones et al. (2010)
	Water stream use and management								Based on Maendly & Humbert (2010) and Humbert &Maendly (2008)

[1] Version 3.0 of LIME, which is currently not documented in English but already available, includes, among other, a water use characterisation model, and focuses on global coverage for many impact categories.
[2] Characterisation Factor
[3] Years of Life Lost (actual unit is [y])
[4] Normalized EXtinction of species
[5] Disability Adjusted Life Years (actual unit is [y])
[6] Potentially Disappeared Fraction of species (not an actual unit but a fraction of 1)
[7] Expected Increase in Number of Extinct Species
[8] Dry Weight
[9] Global Warming Potential (a measure of infrared radiative forcing in [W*y/m2] or in CO2-eq if normalised to CO2)
[10] Intergovernmental Panel on Climate Change
[11] Global Temperature Potential (ratio between global mean surface temperature change at a given future time horizon following an emission of a compound relative to a reference gas (typically CO2)
[12] Ozone Depletion Potential
[13] World Meteorological Organisation
[14] Net Primary Production
[15] PM - Particulate Matter (with diameters up to 2.5μm and 10μm respectively)
[16] Photochemical Ozone Creation Potential
[17] Area Of Protection
[18] World Health Organisation
[19] Volatile Organic Compounds
[20] Non-Methane Volatile Organic Compounds

[21] Sievert, unit of ionizing radiation dose
[22] Reference Dose (US-EPA's acceptable daily oral exposure to the human population likely to be without risk of deleterious effects during a lifetime)
[23] Effective Dose affecting 10% of tested individuals
[24] Effective Dose affecting 50% of tested individuals
[25] Expected Increase in Number of Extinct Species
[26] Chemical Oxygen Demand
[27] Effective Concentration affecting 50% of individuals
[28] Net Primary Productivity

Chapter 4 Exercises

1. Risk Assessment versus LCIA
   An important distinction exists between LCIA and other types of impact analysis, such as traditional risk assessment, i.e. LCIA does not directly assess the impact of chemical releases. Explore this concept further and describe how the data needed to conduct risk assement differ from the data needed for LCIA.
2. Midpoints and Endpoints
   Explain the difference between modeling to the midpoint and modeling to the endpoint. What are the advantages and disadvantages of each approach?
3. Differences in LCIA Models

a) Which LCIA models are global? regional? local? Explain why modeling reflects different spatial levels.

b) Looking at Figure 5, examine the number of impact categories modeled in the three scenarios, and how the labels on the impact categories differ. What are the possible ramifications of having variability in modeling? Are there advantages?

5

Normalization, Grouping and Weighting in Life Cycle Assessment

Abstract

Normalization, grouping and weighting comprise an optional step in the ISO framework for Life Cycle Assessment (LCA). Normalization refers to calculating "the magnitude of category indicator results relative to some reference information." It is intended to provide context and relative magnitude of an impact indicator. Grouping, more rarely described in LCA studies, is "the assignment of impact categories into one or more sets." Weighting is a reflection of value judgments, such as social and political priorities. Typically, weighting factors are applied either to the characterization indicator results or to their normalized version. There is a variety of approaches for normalization, grouping and weighting. Method choice can significantly influence the results.

This chapter describes how normalization, grouping and weighting are currently applied in LCA. The discussion on the normalization explores how this step can be misleading in interpreting results, as argued in Chapter 19 of the LCA Handbook by Prado, Rogers and Seager.

References from the LCA Handbook

Aims of the chapter

1. Describe the optional elements presented in the ISO framework for conducting LCA as part of interpretation of results.
2. Describe internal and external normalization of inventory and impact results and explore how it can influence interpretation.
3. Describe the process of grouping which involves sorting and ranking.
4. Describe various weighting schemes and how they can be applied to data.

5.1 Introduction

Depending on the goal and scope of a study, the following optional elements may also be implemented in the conduct of a Life Cycle Assessment (LCA):

1. Normalization - calculation of the magnitude of category indicator results relative to reference information;
2. Grouping - sorting with the aim of possibly reducing the number of impact categories, as well as possibly ranking them in order of importance;
3. Weighting - converting and possibly aggregating indicator results across impact categories using numerical factors based on value-choices; data prior to weighting should remain available; and
4. Data Quality Analysis - developing a better understanding of the reliability of the indicator results in the impact assessment profile.

As shown in Figure 5.1, the optional life cycle impact assessment (LCIA) elements may use information from outside the LCA framework (ISO 2006). The use of such information should be explained and the explanation should be reported. The application and use of normalization, grouping and weighting methods shall be consistent with the goal and scope of the LCA and fully transparent. All methods and calculations used shall be documented to provide transparency.

Normalization refers to calculating "the magnitude of the category indicator results relative to some reference information." That is, normalization provides context and indicates the relative magnitude of an impact indicator. However, deciding on appropriate normalization methods continues to be an area of controversy. It is an optional step for ISO, and indeed, many LCIA studies stop at characterization.

The reference information for normalization is, in most cases, that total impact in a certain region in a certain time period, e.g., in the country of decision in one year. Normalization helps "to understand better the relative magnitude for each indicator result." Without normalization, the indicator results are in quite different units, e.g., kg CO_2-equivalent for climate change and MJ primary energy for fossil energy depletion. To put these results in perspective, the normalization expresses them as a share of the total impact size in the region. Arbitrary differences due to a choice of units disappear, and it becomes clear to which impact category a product contributes relatively much. The units of the normalize indicator results are equal; nevertheless, such numbers cannot meaningfully be added because the severity of the different impact

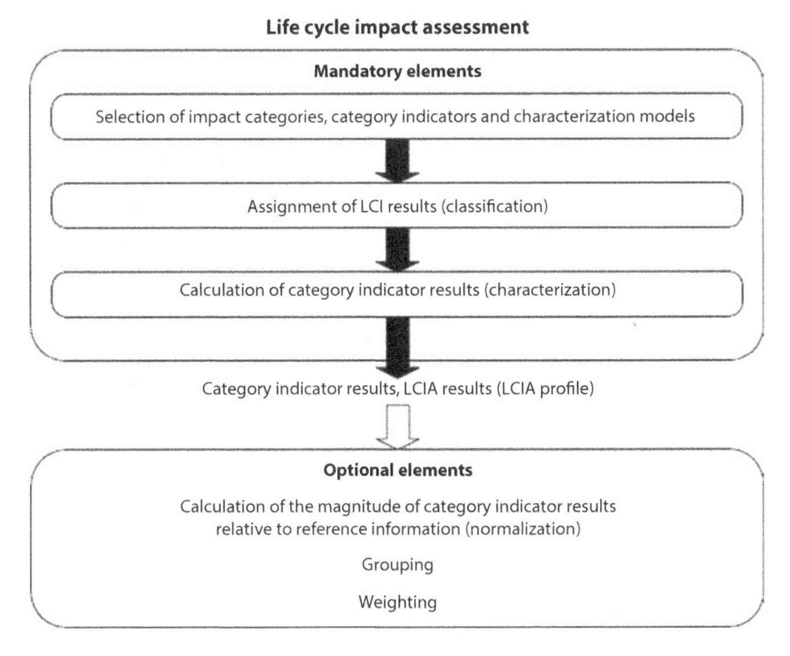

Figure 5.1 Elements of the LCIA phase are mandatory or optional, depending on the goal of the study. This excerpt is from ISO 14040:2006, Figure 4, page 15 with the permission of ANSI on behalf of ISO. © ISO 2015 - All rights reserved.

categories has not yet been accounted for. This can be done in the weighting step; see below. Normalization fulfills several functions: it provides insight into the meaning of the impact indicator results, it helps to check for errors, and it prepares for a possible weighting step.

Grouping is another optional step, although it is seldom seen in LCA studies. ISO defines it as "the assignment of impact categories into one or more sets." ISO mentions two ways:

- Sorting (on a nominal basis, like global/regional/local);
- Ranking (on an ordinal basis, like high/medium/low priority).

Weighting, like characterization, converts and aggregates, but while characterization does so for the LCI results, weighting starts with the characterization (or normalization) results. Typically, weighting factors are applied, either to the characterization indicator results, or to their normalized version. As reflections of value judgements, such as social and political priorities, weighting factors can be obtained several ways. Weighting typically produces one final number, by means of:

$$W = \sum_{c} WF_c \times I_c$$

where I_c again symbolizes the impact score (or normalized impact score) for impact category c, WF_c the weighting factor for this impact category, and W the weighted result. Well-known examples of such weighted results are the eco-indicator and the ELU (environmental load unit).

5.2 Current Practice of Normalization and Weighting in LCIA

The valuation or interpretation stage in LCIA is composed of normalization and weighting, and it helps convey the results of an LCIA study to stakeholders and decision-makers. The results of an LCIA study prior to valuation show the different performances of the alternatives in several impact categories. For example, the performances of a set of products in categories like carbon emissions, water use, and energy requirements. It is difficult to judge the overall environmental performance of alternatives based on multiple criteria with incommensurate units (e.g., tons of CO_2, gallons of water, and kWh).

In practice, when comparing the environmental impacts associated with alternatives, it is rare to find an alternative that outperforms the rest in all impact categories. In fact, most of the time products perform differently in all impact categories, which make normalization and weighting instrumental steps in comparative LCAs. The purpose of normalization is to convert the different units of the impact categories into one dimensionless unit for easier comparison (Bare 2010, De Benedetto and Klemes 2009, Bare *et al* 2006 and, Pennington 2004). Normalization provides context and adds significance to the results. However, deciding on appropriate normalization methods is still an area of controversy (Bare 2010).

After normalization, weighting reflects the relative importance of environmental impacts according to the stakeholders and the decision maker's preferences and values (Seppala *et al* 2002). The weighting process helps to simplify tradeoffs when dealing with competing alternatives and opposing values within the panel of decision makers. For example, a stakeholder might value global warming over ozone depletion. Weighting allows for impacts to be aggregated into a single score for easier evaluation, according to appropriate preferences. However, weights are inherently subjective and can vary depending on culture, political views, gender, demographics, and professional opinion of stakeholders. Consequently, single-score results are criticized by some practitioners. While it is true that other aspects of LCA are also subjective, like the selection of impact categories, Schmidt and Sullivan (2002) make a distinction between choices based on values and choices based on technical assumptions. Therefore, weighting and normalization are categorized as optional steps by the ISO standards.

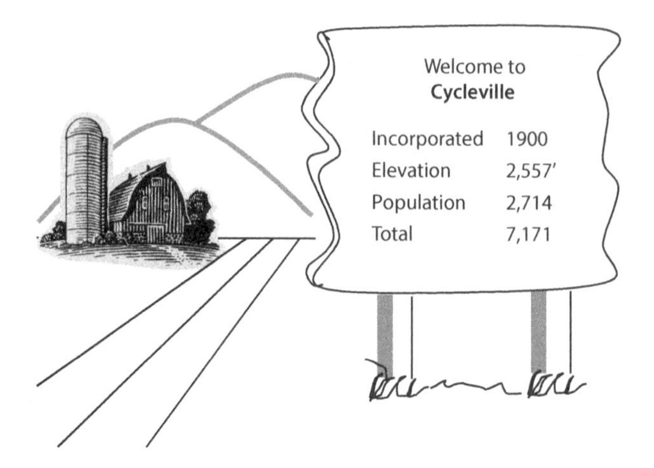

Figure 5.2 Improper Summing of Results to Arrive at a Total.

Current research in LCIA deals primarily with impact categories and characterization factors, and pays little attention to normalization practices. Reap *et al* (2008) perform a survey of major problems in LCA which highlights issues in impact categories and characterization factors, such as spatial variation, local uniqueness, environmental dynamics, and decision time horizon. Bare (2010) mentions termination points (inventory, midpoint, and endpoint) as one of the main research needs in LCIA, and mentions normalization only with respect to the need for more comprehensive external normalization reference databases that report the total amount of emissions in a specific reference system (e.g., total carbon emissions in the US, or total NO_x in the state of California - e.g., Finnveden *et al* 2009).

5.3 Principles of External Normalization

External normalization relates the results of an LCIA study to an external database or normalization reference, thus the results are in terms of a fraction of a broader reference, like total regional or national emissions. External normalization relies on information outside the study and is intended to show the significance of a result relative to a chosen region or reference system (Norris, 2001). By contrast, internal normalization utilizes values within the study and shows the relative significance of an impact with regards to the other competing alternatives.

For example, external normalization relates the carbon emissions of products to the region's total carbon emissions, and internal normalization provides the significance of the product's carbon emission relative to the amount of emissions of the other competing alternatives. Thus, external normalization uses an absolute scale, and internal normalization uses a relative scale (although it can be argued that the "absolute scale" is also relative because it comes from an ideal which is relative by nature -- Saaty 2006).

External normalization is a normative concept based in utility theory which assumes transitivity (Seppala *et al* 2002). Utility theory assigns a number value (or utility) to each alternative with the implicit goal of utility maximization (Fishburn 1970). Thus, an alternative with the greater utility is preferred to lesser. Transitivity requires that when alternative A is preferred over B, and B is preferred over C, then A must be preferred over C (Edwards 1954). Utility theory *rates* alternatives with respect to an *absolute* scale (Saaty 2006). In the case of external normalization in LCIA, the absolute scale is the database of total regional, national or global impacts. Mathematically external normalization is done by dividing the characterized result of each impact category by the value of the normalization reference system:

$$N_i = S_i / A_i$$

Where N is the normalized value for impact category *i*, S is the characterized impact and A is the normalization reference value from an external database (Bare *et al* 2006). The rating of each alternative is independent of each other and it is not subject to change if other alternatives are added or removed (Vargas 1994, Saaty 2006). Therefore, rating in external normalization is transitive. However, not all rational decisions follow a transitive pattern (Vargas 1986). For example, consider the intransitive order of the rock-paper-scissors game: rock beats scissors, scissors beats paper, and paper beats rock. In this case, there is no dominant winning strategy. In fact May (1954) mentions multiple examples that violate the principle of transitivity and shows how intransitivity arises when choosing alternatives with conflicting criteria.

5.4 Issues with External Normalization

External normalization gives context to the characterized results and places different criteria in common terms. However, there are severe disadvantages and fundamental issues that come with applying external normalization to comparative LCAs.

5.5 Inherent Data Gaps

Utilizing external normalization references introduces additional uncertainty to the study because of the lack of consensus in data (Bare *et al* 2006). Any overestimation or underestimation in the external normalization references can have a significant impact in the results (Heijungs *et al* 2007). For instance, a lack of emission data in the NR yields a normalized result that is too high. Such bias is especially problematic when comparing alternatives (White and Clark 2010). Studies dedicated to the reduction of bias in normalization are often concerned with methods for filling data gaps (Bare 2010, White and Carty 2010, Finnveden 2009, Heijungs *et al* 2007). Addressing data gaps is resource intensive and time consuming (White and Carty 2010), and such efforts can prove to be impractical for comparative LCIA studies. Even a comprehensive database can lead to biased results because of fundamental issues such as: risk of masking salient aspects, compensation, boundary issues and discrepancy between different databases.

5.6 Masking Salient Aspects

In external normalization, impact categories with large annual per capita values (e.g., eutrophication) yield small normalized results, as opposed to impact categories with relatively small annual per capita values (e.g., ozone depletion), which yield large normalized values. According to White and Carty (2010) this phenomenon is referred to as "inverse proportionality" and can lead to confusion and counterproductive actions. The bias introduced by external normalization can be so high as to completely exceed the effects of weighting (Rogers and Seager 2009). For example, Figure 5.3 shows this bias by applying six different weight sets to a normalized data, but obtaining the same rank ordering of alternatives in each case. The overall environmental scores change in magnitude, but their ranking remains the same. This shows that the outcome of a comparative LCA study can be independent of stakeholder values' and completely driven by normalization. In Figure 5.3, the weights for HHCR range from zero in long term users, to 61% in short term LCA experts. Similarly, other impact categories like GW the weights range from 9% to 92%, and FFD ranges from 2% to 28%.

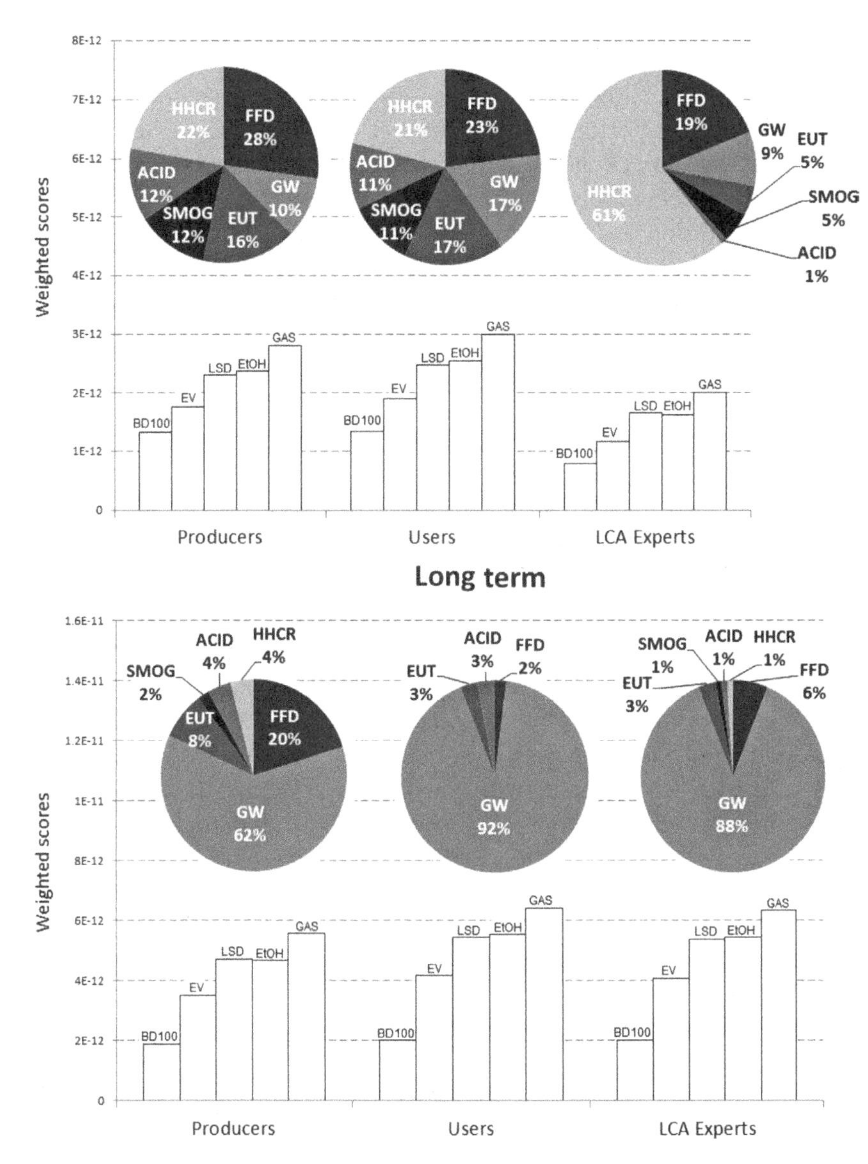

Figure 5.3 Weighted Scores for various transportation fuels: 100% Biodiesel (BD100), Electrical Vehicle (EV), Low Sulfur Diesel (LSD), Ethanol (EtOH), and Gasoline (GAS). The scores are according to a weight set given by Producers, Users and Experts for Short and Long Term impacts. The criteria evaluated were Fossil Fuel depletion (FF), Global Warming (GW), Smog (SMOG), Acidification (ACID), Eutrophication (EUT), and human health criteria air pollutants (HHCR). (Adapted from Rogers and Seager 2009).

5.7 Compensation

External normalization uses utility functions to aggregate values into a single number (Seppala *et al* 2002) and allows for a product's poor performance in one category to be compensated by a good performance in another category. Thus, external normalization is fully compensatory (Rogers 2008). However, compensation is problematic when dealing with environmental decisions because it represents a weak sustainability perspective, to the exclusion of strong sustainability (Rowley and Peters 2009). For example, given a product's outstanding performance in a single category, it is possible that it can offset its poor performance in several others. However, the strong sustainability view rejects such unlimited substitution for pragmatic as well as ideological reasons (Ayers *et al* 1998)

5.8 Spatial Boundaries and Time Frames

Because environmental data is often reported by federal agencies, normalization reference data is typically compiled on a national basis. However, not all environmental impacts have national effects (Bare and Gloria, 2006). For instance, smog has a more localized effect than global warming. Thus, it is possible that impacts outside the reference area will not be accounted for (Heijungs *et al* 2007). Similar to the spatial boundary issues, different processes and products generate emissions over different time periods. Since most normalization references exist on an annual basis, external normalization becomes problematic when dealing with emissions outside this time frame (Finnveden *et al* 2009). For example, landfilling continues to generate emissions even after decades of storage.

5.9 Divergence in Databases

All of the issues combined lead to a great deal of discrepancy between normalization databases. This is clear when different data bases yield significantly different results. White and Clark (2010) offer an example of biases in external normalization. Here, the authors select 800 random materials and processes in the ecoinvent life cycle inventory and utilize two methods of characterization and normalization. The first one uses TRACI[1] characterization factors normalized according to 2000 US per capita values (see chapter appendix "TRACI 2.1 Normalization Factors, US 2008"). The second one uses CML baseline 2001 for characterization factors normalized with CML 1995 database. The results show that the first approach focuses exclusively on human toxicity, human cancer and ecotoxity categories, whereas, the second approach focuses on completely different categories like marine toxicity, freshwater toxicity and fossil fuel depletion.

---

[1] Tool for Assessment and Reduction of Chemical and other environmental Impacts

5.10 Principles of Internal Normalization

Normalization can also be performed internally in a variety of ways, either by division (division by maximum, division by minimum, division by baseline, division by sum), by applying methods from MCDA like the Analytic Hierarchy Process (AHP -- Saaty 1980), or outranking (Behzadian *et al* 2010, Figueira *et al* 2005). Internal normalization is descriptive rather than normative approach, and gives a *ranking* to alternatives that is dependent on other alternatives, rather than a rating. The ranking is based on a relative scale, and it can change when the number of alternatives changes. Relative scales can result in rank reversal (Saaty 2004), but only when relevant alternatives are introduced or removed from the analysis (Harker and Vargas 1990). Rank reversal occurs when the addition or removal of one alternative causes the rank of other alternatives to change. For example, consider that alternative A is ranked higher than Alternative B, but once alternative C is introduced, B becomes the highest ranked alternative followed by A, then C. As opposed to normative approaches, descriptive approaches allow and accept intransitive preferences, thus rank is not always preserved. In fact, the notion of rank preservation or rank invariance principle (Vargas 1994), is a normative concept stating that the ranking of alternatives should remain the same regardless of the number of alternatives introduced or deleted. According to this theory, rank is to be preserved even if there is new information added to the problem. The fact that rank reversal is a real life occurrence is not addressed by the ideal of rank preservation. Rank reversal remains a highly controversial subject within the normative and descriptive communities (Harker and Vargas 1987, Harker and Vargas 1990, Vargas 1994, Erdogmus *et al* 2006, Dyer 1990, Schenkerman 1994), also referred to as classical and naturalistic approaches respectively (Hersh 1999). In LCIA, rank reversal from internal normalization is not well analyzed and understood. Instead, it has been automatically discarded as inappropriate without any further consideration. Initially, LCIA studies applied internal normalization but because of criticisms due to the rank reversal phenomenon, and to ensure congruency in the valuation stage, external normalization became the common practice (Bare 2010, Wang and Elhagi 2006, Norris 2001). Nevertheless, deciding on appropriate normalization guidelines is still an area of controversy (Bare 2010).

5.11 Compensatory Methods

Internal normalization by maximum is a method in which the values of all alternatives in each category are divided by the maximum value in that category prior to weighting. For example, if three alternatives having lead emissions of 2, 4, and 10 mg each were to be normalized, the values will be normalized with respect to the alternative with the highest lead emissions (10 mg of Pb). Thus, it yields dimensionless normalized results of 0.2, 0.4, and 1 respectively. Likewise, internal normalization by minimum would yield 1, 2 and 5. Internal normalization by a baseline, divides the values in the category by the selected baseline alternative. An issue with this method is that it may lead to a division by zero for nonexistent flows (Norris 2001). Division by sum normalization divides the attributes in each category by the sum of the category (Norris and Marshal

1995). A drawback from this method is that it can yield biased results when most values are closer to the top or bottom of the range (Norris 2001). Although these methods do not have the some of the issues of external normalization, internal normalization by means of division still allows for full compensation between categories. This feature leads to an unsatisfactory framework for environmental type decisions where tradeoffs between criteria (e.g., water quality and air quality) are undesirable.

The AHP method was developed by Saaty (1980) with the realization that humans are more capable of making relative judgments over absolute judgments (Linkov *et al* 2007). The AHP uses pair wise comparisons between attributes of two alternatives at a time, and asks questions such as "How much more important is one attribute over the other?" For example, "How much more important is water quality over air quality?" Decision makers are then asked to assign a value from a 0 to 9 scale, where 0 means equally important and 9 means extremely more important. The verbal mediation in the 0-9 scale helps decision makers translate fuzzy judgment into number values (Norris and Marshall 1995). After the pair wise comparisons, an eigenvector analysis yields weights. Once the decision makers assign a value to their preferences and their respective weights calculated, the alternative with the highest overall ranking is said to be the preferred alternative. Although AHP is also a complete method of aggregation that allows for full compensation, it is an intuitive and flexible tool that can deal with tangible and intangible criteria (Ramanathan 2001, Erdogmus *et al* 2006). Nevertheless, AHP is limited in some respects (Macharis *et al* 2004).

5.12 Partially Compensatory Methods

Alternate methods of internal normalization performing outranking such as PROMETHEE (Preference Ranking Organization Method of Enrichment Evaluation) and ELECTRE (ELimination and Et Choice Translating REality), specifically ELECTRE III and PROMETHEE I, II, are advantageous for environmental problems. These methods are partially compensatory, allow for easier value elicitation, and can work with partially quantitative data (Geldermann and Schobel 2011). Outranking judges alternatives with regard to each other on each criterion, provided there is enough evidence to judge one alternative to outrank another (Loken 2007).

There are two main steps to these methods: one involves the normalization process by means of pair wise comparisons, and the second is the process of producing the ranking of alternatives. Both, ELECTRE III and PROMETHEE I and II require a preference function (Figure 5.4) with preference (p) and indifference (q) thresholds. The preference threshold (p) is the smallest deviation between two alternatives considered significant, or enough to be preferred, and the indifference threshold (q) is the largest deviation considered negligible (Brans and Mareschal 2005). Thresholds can be selected arbitrarily (Linkov *et al* 2007) or based on the uncertainty of a given criteria (Rogers and Bruen 1998). Preference values are real numbers between 0 and 1, where 1 is strict preference and 0 is indifference. A weak preference of one alternative over another alternative results in an interpolated preference value between 0 and 1.

After gathering the preference indices for each pair wise comparison, the preference indices for each alternative are aggregated along with the weights. The weights

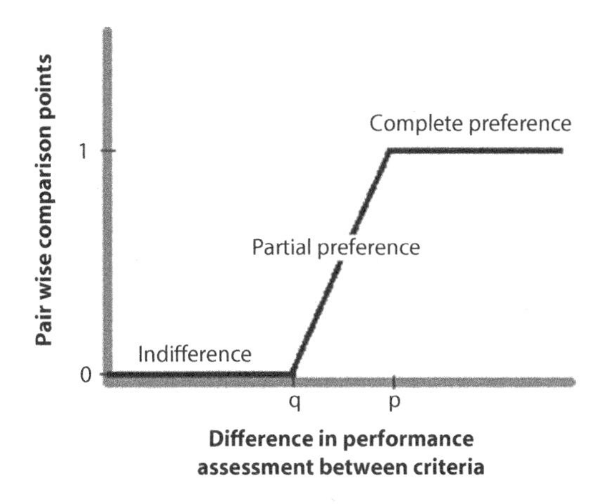

Figure 5.4 Linear Preference Function.

are specific of each impact category, and they reflect the importance of the category as assigned by decision makers. Finally, alternatives are ranked depending on their overall score. The decision making process is an iterative process, and it is not meant to provide an absolute single answer. Instead, it is intended to help decision makers better understand the problem and organize their judgment (Seager *et al* 2006).

Compared to ELECTRE, the calculation procedure in PROMETHEE is more transparent and easier for decision makers to understand (Seager *et al* 2006). It is important for decision makers to understand the methodology so they feel comfortable and trust the recommendations otherwise the decision analysis is meaningless. For example, sometimes the ELECTRE method seems as a "black box" and it is unsatisfactory for decision makers (Loken 2007). PROMETHEE avoids full compensation between criteria, deals with partial quantitative data, and it is easily understood by decision makers. However, PROMETHEE still relies upon point estimates for inputs with no uncertainty. In environmental decisions, uncertainty must be considered because the precise information is not always available within analytic time frames (Hersh 1999). Specifically, there is a need for methods that can investigate the effects of changing input parameters and weights (Hersh 1999). Recently, there have been modified versions of PROMETHEE that allow for uncertainty in the inputs and weights (Rogers 2008, Canis *et al* 2010, Tylock *et al* 2011). These methods utilize Monte Carlo analysis to explore a range of inputs, and allow uncertainty in the input parameters (Lahdelma *et al* 1998). Thus, it is possible to perform an analysis with basic information at an early stage of alternative development or where quantitative performance is difficult to obtain (Seager *et al* 2006).

5.13 Weighting

Weights can be obtained a number of ways (Wang *et al* 2009), but typically are represented as a single vector for easier evaluation. Single-score results are problematic because they lead to an extreme simplification of problem, and lose important information (Brans and Mareschal 2005). Appropriate methods should include sensitivity to weighting analysis (Brans and Mareschal 2005, Hersh 1999, Rogers and Bruen

1998). In fact, there are studies that explore the entire weight set by means of Monte Carlo simulations, resulting in a probabilistic instead of absolute ranking of alternatives (Lahdelma and Salminen 2001, Rogers *et al* 2008).

Norris (2001) exemplifies the dominant views of normalization in LCA, which prefer external normalization and weighting. To prove the point, Norris (2001) presents a multi-alternative, multi-criteria problem normalized internally by division-by-maximum and weighted with single weights. There are two instances in which, according to the paper, the results are debatable. The first example shows that the results are insensitive to changes in magnitude, and the second example shows a case of rank reversal. While Norris (2001) rejects these results as "absurd" without any further analysis, the following sections discuss both examples from a descriptive, rather than normative perspective.

Figure 5.5 presents the example from Norris (2001) in which two alternatives, A and B, are evaluated in three weighted categories: Global Warming, Acidification, and Human Toxics. Alternative B has a higher performance assessment in Acidification and Human Toxics, and Alternative A performs better in the most significant category, Global Warming. After division-by-maximum normalization and external weighting in Figure 5.5b, Alternative A has a lower overall score which means A is preferred to B. (In this case, the score is associated with environmental impact, thus a lower score is better). Figure 5.5b shows the contribution of each category in the overall score. Alternative A has an overall score of 8.5 and Alternative B has a score of 13.1. Although A has a higher score in Human Toxics and Acidification, its score in Global Warming, which was given the highest weight, is significantly lower.

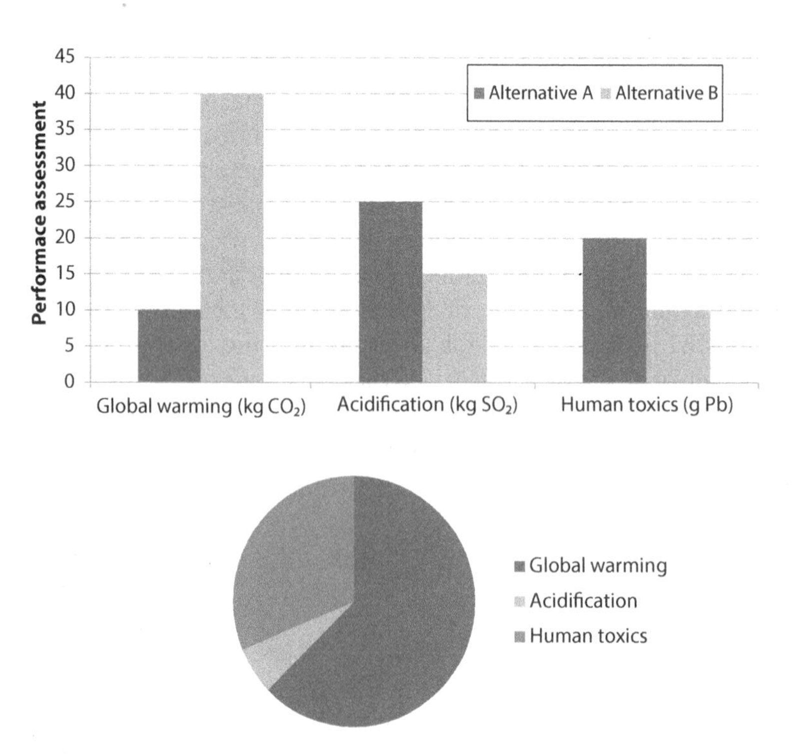

Figure 5.5a Performance Assessment of Alternatives A and B in three categories, note that each category is measured in different units. (Below): Assigned criteria weights (adapted from Norris 2001).

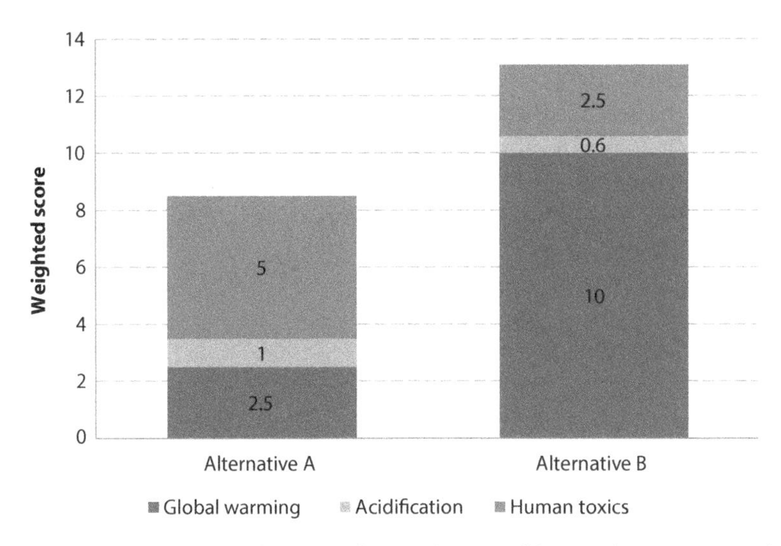

Figure 5.5b Overall weighted score after internal normalization of division by maximum and external single-value weighting for Alternatives A and B (adapted from Norris 2001).

5.14 Multi-Criteria Decision Making

Since comparative LCAs are multi-criteria decision type problems that involve decision makers (policy makers, public, and stakeholders), multiple criteria (e.g., global warming, eutrophication, human toxics, and acidification) and multiple competing alternatives (i.e., different products, policies or services), they can benefit from borrowing tools from decision analysis methods such as Multi-Criteria Decision Analysis (MCDA) to help structure the valuation phase (Rogers *et al* 2008, Rogers and Seager 2009, Jeswani *et al* 2010, Hanandeh and El Zein 2010, Le Teno and Mareschal 1998, Basson and Petrie 2004, Seager *et al* 2008, Benoit and Rousseaux, 2003, Elghali *et al* 2008, Rowley and Shiels 2011, Rowley and Peters 2009, Dorini *et al* 2011). MCDA refers to a variety of methods developed to help decision makers organize and synthesize information to select an alternative among competing options (Loken 2007). The methods are not intended to make actual decisions, instead they are intended to guide the decision making process in a dynamic and iterative manner (Hersh 1999, Seager *et al* 2006). MCDA methods are capable of handling complex decision problems with multiple, conflicting criteria with incommensurate units (Hanandeh and El-Zein 2010, Wang *et al* 2009).

Furthermore, MCDA methods are adequate to sustainability problems because they can integrate environmental, economic and social values (Jeswani *et al* 2010).

There are two main types of MCDA methods that apply to comparative LCAs (Rowley and Peters 2009, Boufateh *et al* 2011). There are normative methods based on the Multi Attribute Utility Theory (MAUT), and descriptive methods such as outranking. MAUT methods are used the most, despite their highly compensatory nature, mathematical complexity, and resource intensity (Seager *et al* 2006). Compensability is a fundamental characteristic of MCDA methods and it refers to the possibility of offsetting poor performance in one aspect of a problem with good performance in another (e.g., clean air makes up for contaminated water, or large profits make up for the loss of ecosystem habitat). Fully compensatory methods are undesirable for environmental problems because they represent an exclusively weak sustainability perspective where different forms of capital (financial, human, and ecological) are considered substitutable (Rowley and Peters 2009). By contrast, outranking methods avoid full compensation and are easier for decision makers to understand (Loken 2007, Benoit and Rousseaux 2003).

References

Basson L and Petrie JG (2004) An Integrated Approach for the Management of Uncertainty in Decision Making Supported by LCA-Based Environmental Performance Information, in Complexity and Integrated Resources Management. Proceedings of the 2nd Biennial Meeting of the International Environmental Modeling and Software Society (iEMSs), edited by C. Pahl-Wostl *et al*, Manno, Switzerland: International Environmental Modeling and Software Society.

Benoit V and Rousseaux P (2003) Aid for Aggregating the Impacts in Life Cycle Assessment *Int J Life Cycle Assess* 8(2):74-82.

Boufateh I, Perwuelz A, and Rabenasolo B (2011) Multiple Criteria Decision-Making for Environmental Impacts Optimization *International Journal Business Performance and Supply Chain Modeling* 3(1):28-42.

Dorini G, Kapelan Z, and Azapagic A (2011) Managing Uncertainty in Multiple-Criteria Decision Making Related to Sustainability Assessment *Clean Techn Environ Policy* 13:133-139.

Elghali L, Clift R, Begg KG, and McLaren S (2008) Decision Support Methodology for Complex Contexts, in Proceedings of the Institution of Civil Engineers, Engineering Sustainability, Vol. 161, pp. 7-22.

Hanandeh AE and El-Zein A (2010) The Development and Application of Multi-Criteria Decision-Making Tool with Consideration of Uncertainty: The Selection of a Management Strategy for the Bio-Degradable Fraction in the Municipal Solid Waste, *Bioresource Technology* Vol 101:555-561.

Hersh MA (1999) Sustainability Decision Making: The Role of Decision Support Systems, IEEE Transactions on Systems, Man, and Cybernetics- Part C: Applications and Reviews, Vol. 29, No. 3, pp. 395-408.

ISO (2006) ISO 14040 Environmental Management – Life Cycle Assessment – Principles and Framework, International Standard, International Organization for Standardization, Geneva, Switzerland.

Jeswani HK, Azapagic A, Schepelmann P and Ritthoff M (2010) Options for Broadening and Deepening LCA Approaches *J Clean Prod* Vol. 18, pp. 120-127.

Loken E (2007) Use of Multicriteria Decision Analysis Methods for Energy Planning Problems, *Renewable and Sustainable Energy Reviews* Vol. 11:1584-1595.

Norris GA (2001) The Requirement for Congruence in Normalization *Int J Life Cycle Assess* 6(2):85-88.

Rogers K and Bruen M (1998) Choosing Realistic Values of Indifference, Preference and Veto Thresholds for Use with Environmental Criteria within ELECTRE *European J of Operational Research* 107:542-551.

Rogers K and Seager TP (2009) Environmental Decision-Making Using Life Cycle Impact Assessment and Stochastic Multiattribute Decision Analysis: A Case Study on Alternative Transportation Fuels *Environ SciTechnol* 43(6):1718–1723.

Rogers K, Seager TP and Linkov I (2008) Multicriteria Decision Analysis and Life Cycle Assessment, in I Linkov *et al* (eds) Real-Time and Deliberative Decision Making. Springer Science, pp. 305-314.

Rowley HV and Peters G (2009) Multi-Criteria Methods for the Aggregation of Life Cycle Impacts, in Proceedings of Sixth Australian Conference on Life Cycle Assessment, Australian Life Cycle Assessment Society, Australia.

Rowley HV and Shiels S (2011) Valuation in LCA: Towards a Best-Practice Approach, in Proceedings of Seventh Australian Conference on Life Cycle Assessment, Australian Life Cycle Assessment Society, Australia.

Ryberg M, Vieira MDM, Zgola M, Bare J and Rosenbaum RK (2014) Updated US and Canadian Normalization Factors for TRACI 2.1 *Clean Techn Environ Policy* 16:329–339.

Saaty T (2006) Rank from Comparisons and from Ratings in the Analytical Hierarchy/Network Processes *European J of Operational Research* 168:557-570.

Seager TP, Rogers SH, Gardner KH, Linkov I and Howarth R (2006) Coupling Public Participation and Expert Judgment for Assessment of Innovative Contaminated Sediment Technologies, Environmental Security and Environmental Management: The role of Risk Assessment, Printed in the Netherlands, pp. 223-244.

Seager TP, Raffaelle RP and Landi J (2008) Sources of Variability and Uncertainty in LCA of Single Wall Carbon Nanotubes for Li-ion Batteries in Electric Vehicles, Electronics and the Environment. IEEE International Symposium, pp. 1-5.

Seppala J, Basson L and Norris G (2002) Decision Analysis Frameworks for Life Cycle Impact Assessment *J Ind Ecol* 5(4):45-68.

Teno JF Le and Mareschal B (1998) An Interval Version of PROMETHEE for the Comparison of Building Products' Design with Ill-Defined Data on Environmental Quality *European Journal of Operational Research* 109:522-529.

Wang JJ, Jing YY, Zhang CF, and Zhao JH (2009) Review on Multi-Criteria Decision Analysis Aid in Sustainable Energy Decision-Making *Renewable and Sustainable Energy Reviews* 13:2263-2278.

Appendix

TRACI 2.1 Normalization Factors, US 2008.

	Impact per year	Impact per person year	Population
Global warming	7.40E+12	2.40E+04	3.08E+08
Ozone depletion	4.90E+07	1.60E-01	3.06E+08
Acidification	2.80E+10	9.10E+01	3.08E+08
Eutrophication	6.60E+09	2.20E+01	3.00E+08
Smog formation	4.20E+11	1.40E+03	3.00E+08
Respiratory effects	7.40E+09	2.40E+01	3.08E+08
Fossil fuel depletion	5.30E+12	1.70E+04	3.12E+08

Table 1 NFs for all impact categories included in TRACI 2.1 calculated using inventories from the US (2008) and US-Canada (2005/2008). The NFs are shown as the geographical areas total potential environmental impact per year, and as the impact per person values.

Impact category	Normalization factors and reference year				
	US 2008		US-CA 2005/2008*		Ratio: US/ US-CA
	Impact per year	Impact per person year	Impact per year	Impact per person year	
Ecotoxicity-non-metals (CTUe)	2.3×10^{10}	7.6×10^{1}	2.5×10^{10}	7.4×10^{1}	1.02
Ecotoxicity-metals (CTUe)	3.3×10^{12}	1.1×10^{4}	3.7×10^{12}	1.1×10^{4}	1.00
Carcinogens-metals (CTUcanc.)	1.7×10^{3}	5.5×10^{-6}	1.7×10^{3}	5.1×10^{-6}	1.08
Carcinogens-metals (CTUcanc.)	1.4×10^{4}	4.5×10^{-5}	1.5×10^{4}	4.3×10^{-5}	1.05
Non-carcinogens-non-metals (CTUnon-canc.)	1.1×10^{4}	3.7×10^{-5}	1.1×10^{4}	3.4×10^{-5}	1.09
Non-carcinogens-metals (CTUcanc.)	3.1×10^{5}	1.0×10^{-3}	3.4×10^{5}	1.0×10^{-3}	1.01
Global warming (kg CO_2 eq)	7.4×10^{12}	2.4×10^{4}	8.0×10^{12}	2.4×10^{4}	1.01
Qzone depletion (kg CFC-11 eq)	4.9×10^{7}	1.6×10^{-1}	4.9×10^{7}	$1.5 \times 10–^{1}$	1.10
Acidification (kg SO_2 eq)	2.8×10^{10}	9.1×10^{1}	3.2×10^{10}	9.5×10^{1}	0.96

Impact category	Normalization factors and reference year				
	US 2008		US-CA 2005/2008*		Ratio: US/ US-CA
	Impact per year	Impact per person year	Impact per year	Impact per person year	
Eutrofication (kg N eq)	6.6×10^9	2.2×10^1	7.00×10^9	2.1×10^1	1.04
Photochemical ozone formation (kg O_3 eq)	4.2×10^{11}	1.4×10^3	4.9×10^{11}	1.5×10^3	0.96
Respiratory effects (kg $PM_{2.5}$ eq)	7.4×10^9	2.4×10^1	1.0×10^{10}	3.0×10^1	0.82
Fossil fuel depletion (MJ surplus)	5.3×10^{12}	1.7×10^4	6.6×10^{12}	1.9×10^4	0.89

* The study used a combined inventory with US and Canada, the reference years are 2008 and 2005 respectively

Ryberg M, Vieira MDM, Zgola M, Bare J, Rosenbaum RK (2014) Updated US and Canadian Normalization Factors for TRACI 2.1 *Clean Technol Environ Policy* 16:329–339.

Chapter 5 Exercises

1. Internal versus External Normalization
   Normalization was defined in the chapter as the "calculation of the magnitude of category indicator results relative to reference information." Explain how internal normalization is done versus external normalization.
2. Optional Steps in ISO
   Why do you think the required steps in the ISO standard stop after characterization and make normalization, grouping and weighting optional?

6

Life Cycle Assessment: Interpretation and Reporting

Abstract
The interpretation step of Life Cycle Assessment (LCA) is where the results of the inventory and impact modeling are analyzed, conclusions are reached, and findings are presented in a transparent manner. It is critical that the reporting of this activity is readily understandable, complete, and consistent with the goal and scope of the study. The aim of life cycle interpretation is to give credibility to the results of the LCA in a way that is useful to the decision-maker.

This chapter explores the key features of LCA interpretation, including the following:

- The use of a systematic procedure to identify, qualify, check, evaluate and present the conclusions based on the results of an LCA or LCI, in order to meet the requirements of the application as described in the goal and scope of the study;
- The use of an iterative procedure both within the interpretative phase and with the other phases of an LCA;
- Maintaining transparency throughout the interpretation phase by clearly stating in the final report any and all preferences, assumptions, or value choices that were used in the assessment or in reporting.

The chapter also deals with how and when to conduct critical review of an LCA, and concludes with a discussion on ISO Type III Environmental Product Declaration (EPD) for documenting the environmental performance of a product using LCA methodology.

References from the LCA Handbook

Aims of the Chapter

1. Help the reader understand the interpretation phase within the LCA framework and its relationship to the other phases, especially the connection to the intended goal of the study.
2. Explore interpretation as a way to systematically identify, quantify, check, and evaluate information from the results of the life cycle inventory and, if conducted, the life cycle impact assessment.
3. Outline the requirements of proper uncertainty and sensitivity analysis to help to prioritize collecting data and making choices.
4. Describe a proper reporting format and ways to present LCA results, including Environmental Product Declarations.

6.1 Introduction

Life Cycle Interpretation is a systematic technique to identify, quantify, check, and evaluate information from the results of the life cycle inventory and, if conducted, the life cycle impact assessment. The results from the inventory analysis and impact assessment are summarized during the interpretation phase. The outcome of the interpretation phase is a set of conclusions and recommendations for the study.

The results from an LCA can be difficult to comprehend because of the vast amount of data, diversity of physical units, use of value judgments, and uncertainty in the parameters. These factors limit its capacity to directly, and transparently, interpret information for decision makers (Canis *et al* 2010, Boufateh *et al* 2011). As a result, many comparative LCA studies stop the assessment after calculation of potential impact indicators (characterized data), leaving the decision makers to confront multi-criteria, multi-stakeholder problems unaided (Rogers *et al* 2008, Rowley and Peters 2009). This can lead to confusion and bias since human cognitive ability to process large amounts of data is limited and subject to systematic flaws (Hertwich and Hammit 2001).

The interpretation phase of LCA entails the evaluation of the results of the inventory analysis along with the results of the impact assessment to aid in the decision making process, whether it is to select the preferred product, improve a process or service, etc. with a clear understanding of the uncertainty and the assumptions used to generate the results. As shown in Figure 6.1, an LCA study is a highly iterative process, so that the LCA practitioner may need to go back to the goal and scope after the preliminary

LCA cannot determine if a product is "sustainable" or "environmentally friendly"

The results can only indicate if product X is "more sustainable" or "more environmentally friendly" than product Y, or that the use phase is the "least sustainable" or "least environmentally friendly" part of the life cycle for product Z, for example.

Impact	Incandescent Lamp	Fluorescent Lamp
Energy Use	60 W	18 W
Climate Change	120,000 kg CO_2-eq	40,000 kg CO_2-eq
Ecotoxicity	320 kg DCB-eq	440 kg DCB-eq
Acidification	45 kg SO_2-eq	21 kg SO_2-eq
Resource Depletion	0.8 kg antimony-eq	0.3 kg antimony-eq

Furthermore, a product can result in lower impact scores in some categories but not in every category across the board. These simplified impact indicator results for two types of lamps (light bulbs) depict how the use of fluorescent lamps instead of incandescent may reduce climate change and acidification impacts, but increases potential for ecotoxicity. Additional interpretation of the results is needed to support the decision making process.

inventory work, to move back from impact assessment to inventory analysis, to have a look at the interpretation in an early stage, etc.

In general, a distinction between procedural and numerical approaches can be drawn:

- Procedural approaches include all types of analyses that deal with the data and results in relation to other sources of information, like expert judgment, reports on similar products, intuition, reputation of data suppliers, and so on.
- Numerical approaches include those approaches that somehow deal with the data that is used during the calculations, without reference to those other sources of information, but as algorithms that use and process the data in different ways, so as to produce different types of "smart" data reduction that provide an indication of reliability, key issues, discernibility, robustness, and so on.

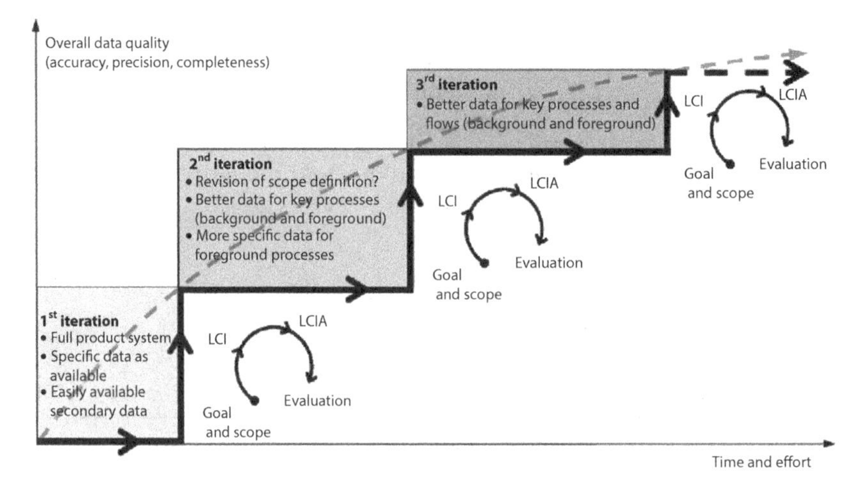

Figure 6.1 Life Cycle Assessment is a highly Iterative Process (JRC 2010).

This distinction helps us understand some important roles of interpretation. On one hand, it is about comparing the data and results with previous findings, and putting the results in the context of decision-making and limitations. On the other hand, it is devoted to a systematic analysis with the help of statistical and other decision-analytic techniques. The latter type may be incorporated in software, and indeed, an increasing number of software packages (see Chapter 3, 3.12 LCA Software) contain options for running Monte Carlo analysis, doing sensitivity analysis, carrying out statistical significance tests, etc. Monte Carlo analysis is discussed later in this chapter.

6.2 LCA Interpretation according to ISO

ISO (2006) defines interpretation as the "phase of life cycle assessment in which the findings of either the inventory analysis or the impact assessment, or both, are evaluated in relation to the defined goal and scope in order to reach conclusions and recommendations." The standard defines two objectives of life cycle interpretation:

1. Analyze results, reach conclusions, explain limitations, and provide recommendations based on the findings of the preceding phases of the LCA, and to report the results of the life cycle interpretation in a transparent manner.
2. Provide a readily understandable, complete, and consistent presentation of the results of an LCA study, in accordance with the goal and scope of the study.

In ISO 14044, the life cycle interpretation phase of an LCA or an LCI study contains several elements, as depicted in Figure 6.2:

- Identification of the significant issues based on the results of the LCI and LCIA phases;

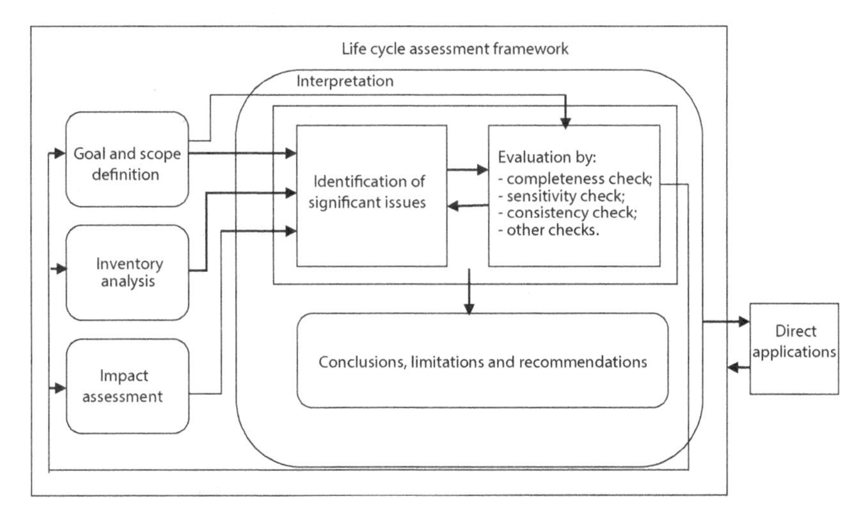

Figure 6.2 Detailed Interpretation Phase and its Relationship to the Other Phases within the LCA Framework. This excerpt is adapted from ISO 14044:2006, Figure 4, page 24 with the permission of ANSI on behalf of ISO. © ISO 2015 - All rights reserved.

- An evaluation that considers:
  - o Completeness check- to ensure that all relevant information and data are available and complete
  - o Sensitivity analysis - to assess the reliability of the final results and conclusions
  - o Consistency check - to determine whether the assumptions, methods and data are consistent with the goal and scope;
- Conclusions, limitations, and recommendations.

Other important elements of interpretation noted in the ISO standard include:

- Appropriateness of the definitions of the system functions, the functional unit and system boundary;
- Limitations identified by the data quality assessment and the sensitivity analysis.

The text of ISO on interpretation is very brief, giving no details on procedures and techniques to be employed. The same applies to most guidebooks on LCA. They mention carrying out an uncertainty analysis, but give no clear guidance on how this should be done.

The iterative nature of the ISO framework shows up in this context. It is especially important to determine that if the results of the impact assessment or the underlying inventory data are incomplete or unacceptable for drawing conclusions and making recommendations, i.e. the uncertainties are too high, then those steps must be repeated until the results can support the original goals of the study. Whenever sensitivity analysis shows that some decisions are crucial, we may go back and do a more refined analysis. In this way, the interpretation helps to prepare for a balanced decision, while helping improve the LCA.

6.3 Uncertainty and Sensitivity Analysis

ISO 14040 describes uncertainty analysis as "a systematic procedure to quantify the uncertainty introduced in the results of a life cycle inventory analysis due to the cumulative effects of model imprecision, input uncertainty and data variability." Potential sources of uncertainty include:

- The data source itself (random or systematic error in measurement and sampling or natural variability),
- Any assumptions and/or calculation use to manipulate data, and
- Aggregating very different data sources or values.

Sensitivity analysis can be used to indicate which parameters are most important to the analysis. Both sensitivity analysis and uncertainty analysis are useful in determining where data quality resources should be directed or redirected. As mentioned in Chapter 3, LCA studies continue to be published without uncertainty or sensitivity analysis. There is of course a psychological argument that a contractor pays for finding out something, not for increasing the doubt. And as many LCA practitioners spend several months on collecting data, it is never pleasant to feel as if efforts were wasted in a last-minute uncertainty analysis. But decision-making obviously means also taking into account the limits of knowledge. Moreover, as discussed before, a proper analysis of uncertainties and sensitivities helps to prioritize the steps earlier on in the framework: collecting data, setting boundaries, making choices.

6.3.1 Uncertainty Analysis

Due to the uncertainty with inventory and impact data, it may not be possible to state that one alternative is better than the others because of the uncertainty in the final results (see Figure 6.3). This does not imply that the effort was pointless or that LCA is not a viable tool for decision making. The LCA process will still improve understanding of the environmental and health impacts associated with each alternative, where they occur (locally, regionally, or globally), and the relative magnitude of each type of impact in comparison to each of the proposed alternatives included in the study. This information more fully reveals the pros and cons of each alternative.

6.3.2 Uncertainty in Impact Models

Typical environmental and human health impact models, such as those used in risk assessment, are based on fate, exposure and effect (that is, where the substance goes once it is emitted to the environment, who is exposed to the substance and how much, and how toxic the substance is). Impact models used in LCA work with a more generalized fate model, with average high or low population densities, average wind, average crop production etc. This leads to higher uncertainties by definition, but allows it to work in the LCA context. Effects on human health are based on data developed in toxicological studies where safety factor are set to determine acceptable daily intake. In general, this approach adds to the uncertainty. A safety factor can be on the order of 100

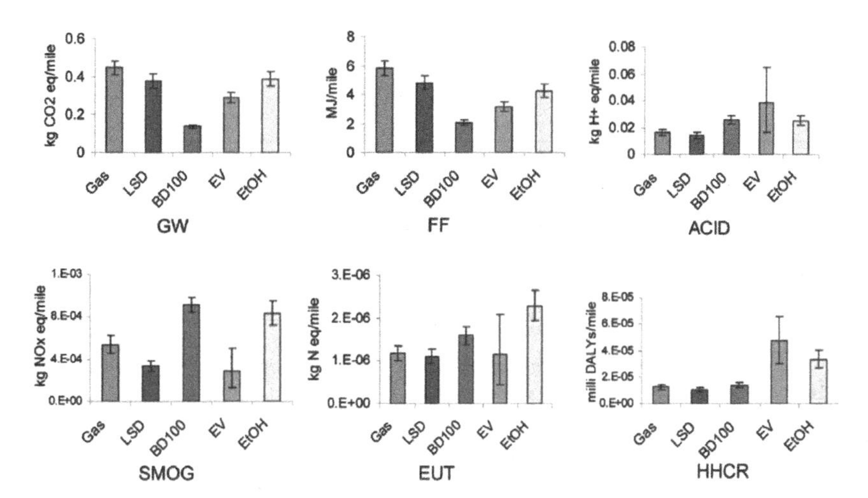

GW: global warming; FF: fossil fuels; ACID: acidification; SMOG: smog formation; EUT: eutrophication; HHCR: human health cancer risk.

Figure 6.3 Showing Uncertainty in Life Cycle Assessment Results for Gasoline (Gas), Low Sulfur Diesel (LSD), Biodiesel (BD100), Electric Vehicle (EV) and Ethanol (EtOH) (Rogers and Seager 2009).

or even 1,000 for substances with high uncertainty, such as heavy metals. So, even if we have a perfect fate model, we still have high uncertainties in the effect model. This type of approach has become all very acceptable.

Different impact models may use different methodological choices, depending on the specific systems being studied, as well as issues that are the focus of increasing interest in the environmental community, such as water use and carbon tracking. In cases where the choice of methodology has a strong influence on the study results and conclusions, the practitioner should justify the reasons for the methodology chosen, and a sensitivity analysis should be conducted to see if an alternative methodological choice produces similar or different results and conclusions. For example, because of the many uncertainties surrounding biomass decomposition in landfills, it is advisable to conduct sensitivity analyses on the carbon storage and releases associated with landfilled biomass products.

6.3.3 Sensitivity Analysis

Sensitivity analysis is done by systematically changing an input parameter and observing the impact on the results. It may be useful in the following situations:

- The analyst does not have a high degree of confidence in an important data source,
- The production system being assessed is highly variable, or
- Data for a particular element are missing or deficient (EPA 1995).

As an example of applying sensitivity analysis to life cycle inventory, consider two cartons that hold an equivalent volume of ice cream. Carton A is cylindrical with a lift-off lid, while Carton B is rectangular with a paperboard flap closure. Carton B can

Table 6.1 Hypothetical Tornado Diagram Worksheet for Product X (EPA 1995).

Model Parameters	Range of Change in Parameter Values[a]	Range of Change in Air Releases (lbs)
Virgin Material	1–10%	1,000–30,000
Energy Consumption	1–10%	1,000–24,000
Recycle Materials	1–10%	1,000–19,000
Water Use	1–10%	7,500–24,000
Transportation	1–10%	5,500–18,500
Packaging	1–10%	5,000–14,000
Water Releases	1–10%	8,500–12,500
Hazardous Waste Releases	1–10%	9,000–11,000
Solid Waste Releases	1–10%	9,750–10,075

[a] The range of change of 1-10% was chosen for example purposes only. In practice, reasonable values specific to the parameter would be more appropriate.

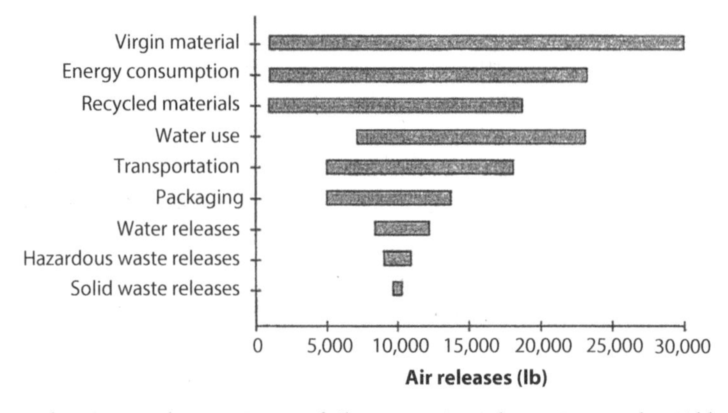

Figure 6.4 Tornado Diagram showing Range of Changes to Air Releases Reported in Table 6.1.

be packed more compactly in a store freezer, so that less shelf space is required for ice cream packaged in Carton B compared to the same quantity of ice cream packaged in Carton A. Because Carton A occupies more freezer space per unit volume of ice cream, Carton A is allocated a larger share of the daily energy requirements for operating the store freezer. However, consumers tend to prefer Carton A's removable lid design over Carton B's paperboard flap closure. If consumer preference for the Carton A design translates into faster sales compared to Carton B, the reduced time in the retail freezer for Carton A may offset its additional freezer shelf space requirement. When consumer behavior is involved, it is advisable to conduct a sensitivity analysis unless data are available to reliably characterize actual consumer behavior.

Such "one-way sensitivity analysis" is useful for evaluating the importance of individual parameters to model results. The sensitivity of individual parameters relative to the system total is determined by calculating the amount of individual parameters would need to change in order for the model results to be altered by a given percentage. Table 6.1

shows how a 1–10% change in input parameters affects air releases. In practice, the range of reasonable values must be evaluated and an appropriate range determined for any each input. Figure 6.4 shows the same results in the format of a tornado diagram.

A SIMPLE BUT NON-LINEAR SYSTEM

A system delivers 1000 kWh of electricity, consisting of only two unit processes, but these processes are dependent on each other.

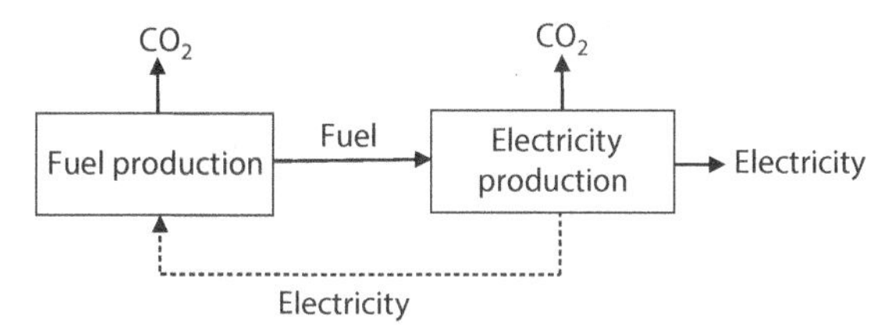

Viewed independently, the life cycle inventory looks like this:

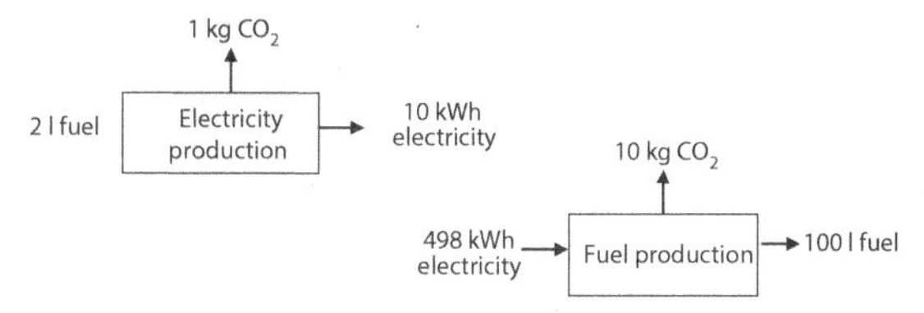

LCA software programs can easily handle systems with such recursive flows. Straightforward mathematics[1] provides an expression for the CO_2-emission of the form:

$$CO_2 = \frac{100 \text{ x } 1000 \text{ x } 1 + 2 \text{ x } 1000 \text{ x } 10}{-2 \text{ x } 498 + 10 \text{ x } 100} = 30$$

The system-wide emission of CO_2 is calculated to be 30 kg. This result is highly dependent on the coefficient 498, the amount of kWh of electricity that is needed to produce 100 liter of fuel. If the 498 is changed by just one unit, to 499, and keeping the other five coefficients unchanged, the CO_2 emissions double to 60 kg. The denominator plays a central role in the extreme sensitivity. In the present form, it is 0.25 (-2 x 498 + 10 x 100), but when 498 is changed into 499, it doubles to 0.5. Thus, even though the formulation of the system is expressed with linear equations, the solution is a non-linear one (Heijungs 2002).

6.3.4 Monte Carlo Simulation

Monte Carlo simulation is a widely used method to perform uncertainty and sensitivity analysis. It uses statistical sampling techniques to approximate the probability of certain outcomes by running multiple trial runs, called simulations, using random variables. Many commercial LCA programs now offer Monte Carlo analysis.

---

[1] For more on the computational structure of LCA modeling, the reader is referred to Heijungs and Suh 2002.

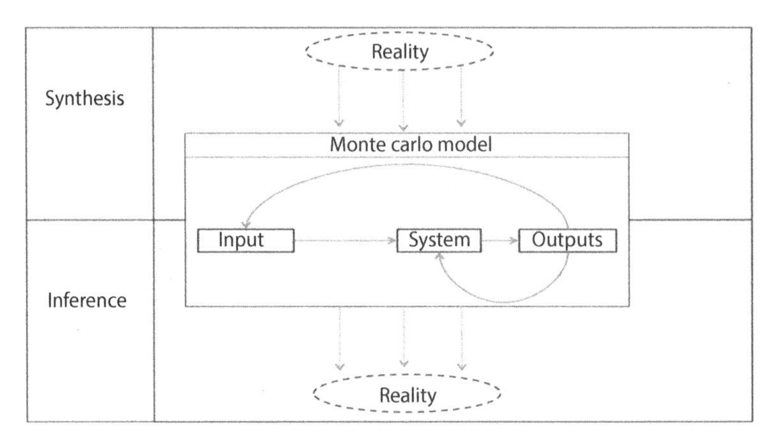

Figure 6.5 Overview of Monte Carlo Analysis to Determine Uncertainty.

Developed in the 1940's, Monte Carlo simulation is a statistical technique in which a quantity is calculated repeatedly, using randomly selected "what-if" scenarios for each calculation. Though the simulation process is internally complex, commercial computer software performs the calculations as a single operation, presenting results in simple graphs and tables. These results approximate the full range of possible outcomes, and the likelihood of each (EPA 1994). This type of model is usually deterministic, meaning that you get the same results no matter how many times you re-calculate. A simulation can typically involve over 10,000 evaluations of the model, a task which in the past was only practical using super computers, but can be done easily today.

6.4 Contribution Analysis

By comparing the contribution of the life cycle stages or groups of processes to the total result, and examining them for relevance, contribution analysis determines which inventory data or impact indicator has the biggest influence[2]. Usually, this is expressed as a percent of the total. Table 6.2 provides a simplified example of the contribution of greenhouse gases (measured in CO_2-equivalents) toward climate change for an incandescent versus a fluorescent lamp (light bulb). These results show that the main contribution of emissions occurs during electricity production, a result of the use of the lamp.

Analysis of the contributions of the different inputs and outputs identifies which processes or life cycle stages are the most impacted (Table 6.3). On this basis, later evaluation can reveal and state the meaning and stability of those findings that then form the bases for conclusions and recommendations. This evaluation may either be qualitative or quantitative.

[2] The investigation into which process or life cycle stage shows the highest impact compared to the total has also been referred to as "dominance analysis" (for example, Baumann and Tillman 2004).

Table 6.2 Contribution Analysis of Greenhouse Gas Emissions Contributing to Climate Change.

Process	Incandescent Lamp	Fluorescent Lamp
Electricity Production	88%	60%
Copper Production	5%	15%
Waste Disposal	2%	10%
Other	5%	15%
Total Climate Change	120,000 kg CO_2-eq	40,000 CO_2-eq

Table 6.3 Analysis of LCI results identifies the inputs and outputs that contribute the most to the processes or life cycle stages (ISO 2000).

LCI	Material Production		Manufacturing		Use		Other		Total	
	kg	%	kg	%	kg	%	kg	%	kg	%
Hard coal	1200	69.6	25	1.5	5400	28.9	-	-	1725	100
CO_2	4500	66.7	100	1.5	2000	29.6	150	2.2	6750	100
NOx	40	44.5	10	11.1	20	22.2	20	22.2	90	100
Phosphates	2.5	8.9	25	89.3	0.5	1.8	-	-	28	100
AOX	0.05	8.2	0.5	82.0	0.01	1.6	0.05	8.2	0.61	100
Solid Waste	15	8.7	150	87.2	2	1.2	5	2.9	172	100
Tailings	1500	85.7	-	-	-	-	250	14.3	1750	100

AOX: absorbable organic halides

According to Annex B of ISO 14049 (2000), Examples of Life Cycle Interpretation, contribution analysis can take one of three forms:
b) *dominance analysis*, in which, by means of statistical tools or other techniques such as quantitative or qualitative ranking (e.g. ABC analysis), remarkable or significant contributions are examined;
c) *influence analysis*, in which the possibility of influencing the environmental issues is examined; or
d) *anomaly assessment*, in which, based on previous experience, unusual or surprising deviations from expected or normal results are observed. This allows a later check and guides improvement assessments.

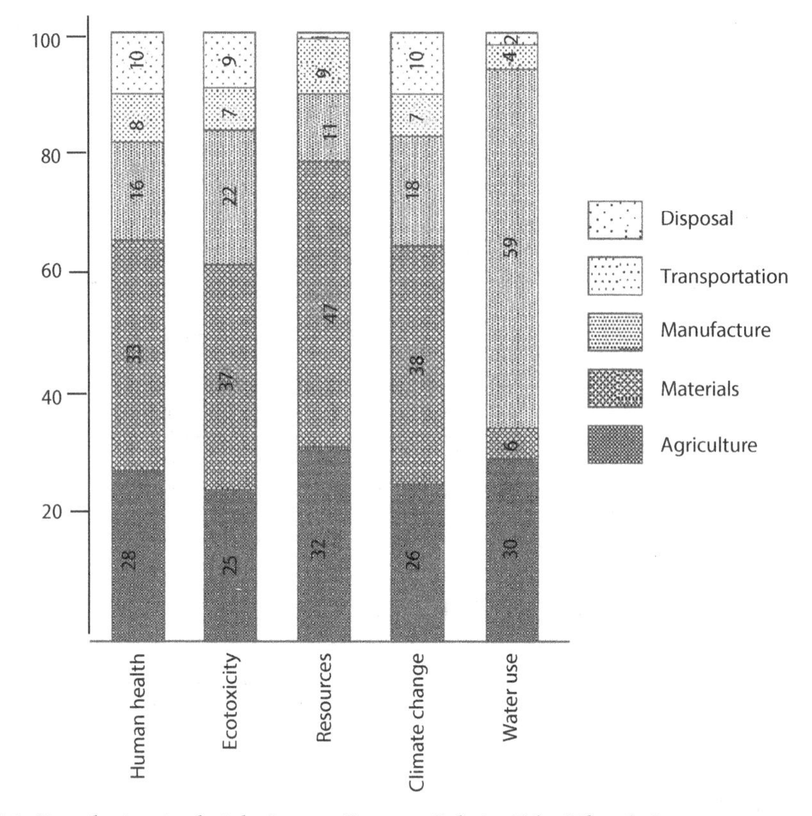

Figure 6.6 Contribution Analysis by Impact Category, Relative % by Lifecycle Stage.

The result of this determination process may also be presented as a matrix, in which the above-mentioned differentiation criteria form the columns, and the inventory inputs and outputs or the category indicator results form the rows.

6.5 Presenting LCIA Results

Impact indicator results from an LCA are presented in many formats. Listing indictor scores in tabular format is the most straightforward way. For a more visual approach, bar charts are commonly used. The results are often normalized to the higher value (as shown in Figure 6.6). Another ways to present LCA results is a "spider diagram," also called a "target plot," an n-dimensional polygon where n is the number of indicators (Figures 6.7–6.8).

Users may be tempted to calculate and compare the areas under each "web." However, this approach does not include a normalization and weighting step (discussed in Chapter 5), so the results would likely be misleading.

6.6 Preparing the Final Report

Once the LCA has been completed, the materials must be assembled into a comprehensive report documenting the study in a clear and organized manner. This will help

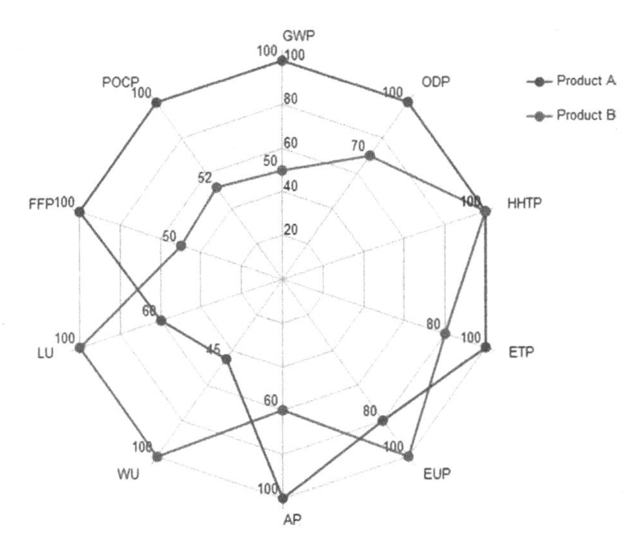

GWP: global warming potential; ODP: ozone depletion potential; HHTP: human health toxicity potential; ETP: ecotoxicity potential; EUP: eutrophication potential; AP: acidification potential; WU: water use; LU: land use; FFP: fossil fuel use potential; POCP: photochemical oxidation creation potential

Figure 6.7 Reporting Impact Indicators in a Spider Diagram (normalized to the product with the highest impact score).

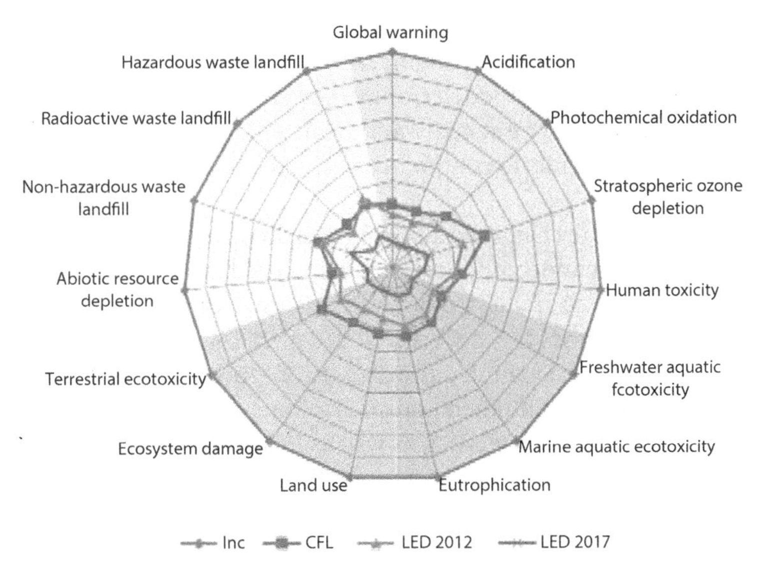

Figure 6.8 Comparing Four Lamps - Incandescent (Inc), Compact Fluorescent (CFL), and Light Emitting Diode (LED) for the present (2012) and future (2017) technology (DOE 2013).

communicate the results of the assessment fairly, completely, and accurately to others interested in the results. The report presents the results, data, methods, assumptions, and limitations in sufficient detail to allow the reader to comprehend the complexities and trade-offs inherent in the LCA study.

If the results will be reported to someone who was not involved in the LCA study, i.e., third-party stakeholders, this report will serve as a reference document and should be provided to them to help prevent any misrepresentation of the results.

The reference document should consist of the following elements (ISO 1997):

1. Administrative Information
   a) Name and address of LCA practitioner (who conducted the LCA study)
   b) Date of report
   c) Other contact information or release information
2. Definition of Goal and Scope
3. Life Cycle Inventory Analysis (data collection and calculation procedures)
4. Life Cycle Impact Assessment (methodology and results of the impact assessment that was performed)
5. Life Cycle Interpretation
   a) Results
   b) Assumptions and limitations
   c) Data quality assessment (including significant findings of uncertainty and sensitivity)
6. Critical Review (internal and external)
   a) Name and affiliation of reviewers
   b) Critical review reports
   c) Responses to recommendations

While conducting the LCA (within both the LCI and LCIA) it is necessary to apply various modeling assumptions and engineering estimates. At times these choices are based on the values held by the modeler, or by the person who commissioned the study. Therefore, every choice must be stated and the impact on the decision clearly communicated within the final results to comprehensively explain conclusions drawn from the data.

Sometimes, appropriate characterization factors do not yet exist for modeling inventory data. For example, nanocomponents are being used in product manufacture at an increasing rate, yet data to model potential human health or ecotoxicity impacts are often lacking. The same goes for the "critical rare metals and metalloids" that are used in increasing amounts in the manufacture of electronic products, such as indium which is in LCD displays for TV, computer, and smartphone screens. Acceptable models for indicating potential resource depletion, as well as toxicity, are still under development. Even though it may not be possible to communicate this information numerically in the impact assessment phase, this is valuable, qualitative information that should be retained and presented in the final report, along with inventory tables and other impact charts. In the end, an LCA report should be viewed as a mechanism to produce complete (as possible) knowledge about the environmental impacts of a product, product, or activity.

Example Contents of an LCA Report[3]

[3] Based on Life Cycle Assessment of Drinking Water Systems: Bottled Water, Tap Water, and Home/Office Delivery (HOD) Water, State of Oregon Department of Environmental Quality, 09-LQ-104, October 22, 2009.

6.7 The Review Process

Critical review has existed almost as long as life cycle assessment (LCA) itself. The intent of reviewing the conduct and reporting of an LCA is to further improve the study, remove serious errors, and deter deceptive practice, especially in the case of comparative studies. Although "peer" review for LCA studies was first proposed in the 1993 SETAC guidelines "A Code of Practice", the first detailed guidance for critical review of LCA was not published until 1997 (Weidema 1997, Klöpffer 1997). The international standards series for LCA, originally developed between 1996 and 2000, and since updated to ISO 14040 and ISO 14044 in 2006, has addressed the issue and makes review a requirement for an LCA study. ISO 14040 describes, among other things, three types of "critical review" which are optional in general, but mandatory "for LCA studies used to make a comparative assertion disclosed to the public." Additionally, since 2006, the minimum number of experts in a "review by interested parties" is now three (including the chair).

Much collective wisdom and experience has been gained with the conduct of reviews in practice. However, specifics on how and when to conduct critical review of LCA studies are still lacking. This has led to the development of additional guidance under ISO TS 14071[4]. The intent of this proposed international technical specification is to provide requirements and guidelines for conducting a critical review and the competencies required (Finkbeiner 2013).

6.7.1 ISO-Defined LCA Review

First, we need to address the terms "critical review" and "verification" which tend to be used interchangeably, even though they are philosophically different (Grahl and Schmincke 2012). "Critical review", like scientific peer review, checks the scientific capacity of a study, and appropriateness of methods and assumptions. In comparison, "verification" is an audit exercise and is the validation of conformance5 against specified requirements, based on an evidence-based approach supported by principles of objectivity and repeatability (see also ISO 19011:2011). Unlike verification, critical review does not rely on objective evidence, but instead relies on expert judgment about LCA methodology and practice. The well-known five criteria presented by ISO 14044 clause 6.1 (see box) for a critical review are not operationalizable in the same way verifiable requirements are.

[4] ISO/PDTS 14071 Environmental management -- Life cycle assessment -- Critical review processes and reviewer competencies: Additional requirements and guidelines to ISO 14044:2006, technical specification under development; 2013-09-2.

[5] The word "conformance" is preferred over "compliance" which implies agreement with some external authority.

ISO 14044 (2006) specifically Identifies five requirements in the critical review process

Clause 6.1 General

The critical review process shall ensure that:

- the methods used to carry out the LCA are consistent with this international standard;
- the methods used to carry out the LCA are scientifically and technically valid;
- the data used are appropriate and reasonable in relation to the goal of the study;
- the interpretations reflect the limitations identified and the goal of the study;
- the study report is transparent and consistent.

ISO 14044 (2006) defines "critical review" as the "process intended to ensure consistency between a life cycle assessment and the principles and requirements of the International Standards on life cycle assessment[6]." Since the ISO standard specifies "critical review", this terminology is used this paper even though the intent of current LCA reviews may be more in line with a checklist-type approach to verify conformance with the ISO standards.

6.7.2 Conduct of an LCA Review

The ISO 14044 standard requires that the scope of the study define whether a critical review is necessary and, if so, how to conduct it; the type of critical review needed; and who would conduct the review, and their level of expertise. The standard is very specific when it comes to review of results that are intended to be used to support a comparative assertion intended to be disclosed to the public. In order to decrease the likelihood of misunderstandings or negative effects on external interested parties, a panel of interested parties shall conduct critical reviews on these types of studies.

Clause 6.2 of ISO 14044 indicates that a critical review may be carried out by a single internal or external expert. In either case, the expert must be independent of the LCA. Under clause 6.3, critical review may also be carried out by a panel of interested parties. In such a case, an external independent expert should be selected by the original study commissioner to act as chairperson of a review panel of at least three members. Based on the goal and scope of the study, the chairperson should select other independent qualified reviewers. This panel can include other interested parties who are affected by conclusions drawn from the LCA, such as government agencies, non-governmental groups, industry competitors and affected industries.

Typically in practice, members of the review team (also known as the panel) bring a combined expertise that offers complete coverage of all aspects involved in conducting an LCA. At a minimum, these specialties must include knowledge in the following areas:

[6] The definition in ISO 14044 for critical review (section 3.45) includes NOTE 1 The principles are described in ISO 14040:2006, 4.1 and NOTE 2 The requirements are described in this International Standard.

- Conformity to the ISO standard at goal and scope definition;
- Data/inventory (LCI) relevant to the industry sector(s) being studied;
- Life cycle impact assessment (LCIA) methods, data and modeling of relevant impact categories; and
- Overall flow and management of the review process.

While each member of the panel reads the whole report, the depth of the critical reading may and should differ. Sometimes, one member of the review team is a technical specialist for the product, industry or production technology which is in the focus of the study; this member is not necessarily an LCA expert, although it is usually better that all members are familiar with the approach (Klöpffer 2005, 2012).

The expert review of LCAs requires a close (as in intimate) and open co-operation between the commissioner of the study, the practitioner, and the reviewer or the review team. All identities are known, unlike the peer review of scientific journals, which are often conducted in a single blind approach (the reviewers are anonymous). LCA critical review is also much more time consuming.

The review statement and review panel report, as well as comments of the expert and any responses to recommendations made by the reviewer or by the panel, are to be included in the LCA report.

6.7.3 Review of Inventory Data

An important aspect concerning quality and credibility of any LCA is access to the underlying inventory data. In fact, the confidentiality of sensitive data is the most important reason for conducting a professional review. ISO 14044 does not directly describe data review; however, draft ISO TS 14071 indicates that the critical review should cover all aspects of an LCA, including data and calculation procedures: "The critical review should cover all aspects of an LCA, including data, calculation procedures for linking the unit processes into product systems, life cycle inventory impact assessment methodologies, characterization factors and calculated LCI and LCIA results."

However, since not all data are publicly available, or they are published in a highly aggregated form, reviewers have to judge data quality and appropriateness, for example by random sampling or tracking of calculations. In order to do so, they have to have access to the data, and the commissioner and practitioner must provide this access. This may be complicated if third parties provide data or are not prepared to reveal confidential issues. Sub-contractors have to be included in data transparency for the review team or at least for one member of the panel responsible for data quality.

6.7.4 Timing the Review

A review can be conducted at various stages of completion of the LCA study (as shown in Figure 6.9). ISO 14044 does not elaborate on this issue. However, draft ISO TS 14071 offers guidance if the review is performed concurrently or at the end of the study (i.e., post-study):

- If the critical review is performed at the end of the LCA study, the process starts when the draft LCA report is provided to the reviewer(s).

- If the critical review is performed concurrently with the study, the process starts as early as the study commissioner and the practitioner decide. The various milestones at which the reviewer(s) may be asked to submit comments and recommendations are:
  - The goal and scope definition;
  - Inventory analysis, including data collection and modelling;
  - Impact assessment;
  - Life cycle interpretation; and
  - Draft LCA report.

The process choice does not change the deliverables of the critical review process. In all cases, the process shall clearly define and document its assessment of choices made by the study practitioners in areas including boundaries, functional unit, data, allocations, indicators, etc.

6.8 Product Category Rules and Environmental Product Declarations

Environmental product claims, which make up a significant proportion of product sustainability claims, take a wide array of forms and may be based on very straightforward, single criterion (such as '% recycled content' or on rather complete studies

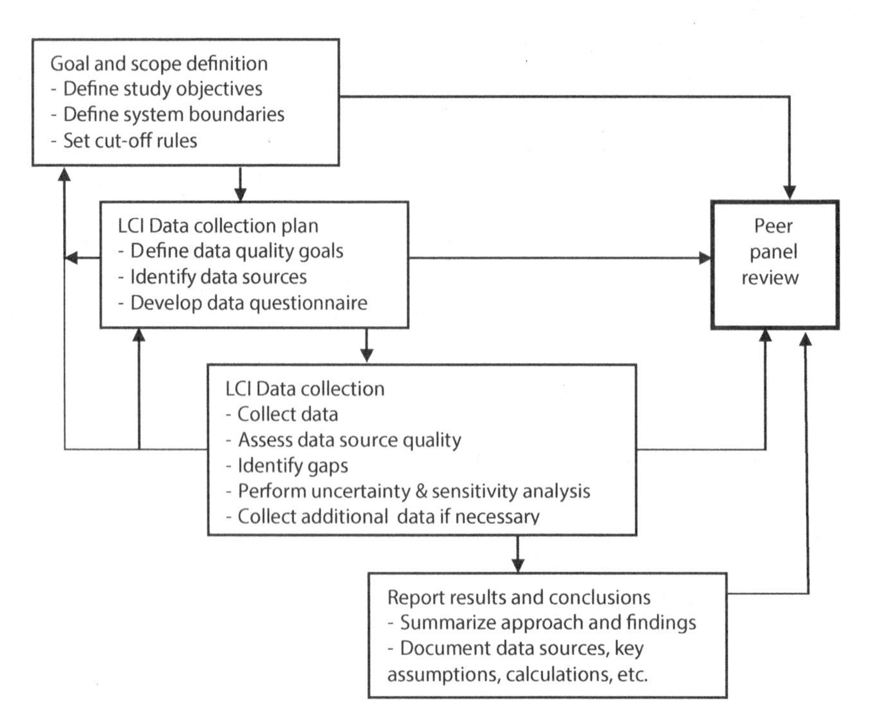

Figure 6.9 Timing the Peer Review.

of multiple environmental impacts of the full life cycle of the product documented in extensive reports. A classification system for voluntary labels has been presented in the ISO 14020 series categorizing environmental product and service claims as Type I (third party certified, specifics in ISO 14024), Type II (self-declared, specifics in ISO 14021), and Type III (third party based on LCA, specifics in ISO 14025). Of these ecolabeling types, Type III, also known as Environmental Product Declarations, are the most closely tied to the LCA methodology.

- Type I, Type II, and Type III claims and their respective relationship with life cycle assessment;
- Product Category Rules & Environmental Product Declarations;
- PCRs in Carbon Footprinting and Product Index development; and
- "Other relevant environmental information" within PCRs informed by other assessment methods including: Water Footprinting, Toxicity Assessment and Ecosystem Services Assessment.

6.8.1 Type III Environmental Product Declarations

ISO 14025 (ISO 2006c) describes a Type III Environmental Product Declaration (EPD) as a document which indicates the environmental performance of a specific product, providing quantified environmental data developed using the LCA methodology set forth in ISO 14040 and 14044 and a predetermined set of rules for the assessment called "Product Category Rules" and, where relevant, additional environmental information (e.g., toxicity in the use phase). ISO 14025 outlines requirements for their content, transparency and verification. EPDs are product- and company-specific and thus owned by the company that develops them.

There are several aspects of this definition worth discussion including: 1. An EPD is a document; 2. An EPD is primarily based on the LCA methodology outlined in ISO 14040 and 14044; 3. An EPD is developed by following a "Product Category Rule"; 4. An EPD can contain information beyond the scope of an LCA, where relevant to that product.

6.8.2 An EPD is a Document

As previously discussed, typically Type I claims appear on a product as a stamp or seal of approval and similarly, Type II claims appear as a short statement calling out one attribute on a product. Contrarily, Type III product declarations are often a multi-page document outlining detailed environmental information. While the format of an EPD can vary based on the intended audience and program goals; generally, they are thought of as communicating to a more "sophisticated" audience of institutional buyers and not the average consumer. Sometimes EPDs are described as nutrition labels including environmental information, however it should be noted that solely a multi-attribute label expressing the life cycle impacts assessment results for a product would not meet all of the required documentation outlined in the ISO standard for an EPD.

6.8.3 An EPD is Primarily Based on LCA

ISO 14025 specifically lists the ISO 14040 series as the basis for data and inventory analysis supporting the development of an EPD and its supporting PCR. This is different from the guidance in ISO 14024 for Type I labels which indicates that they need to be based on life cycle considerations. ISO 14025 specifies that EPDs can include raw data from life cycle inventory results (e.g., fossil fuel or water consumption), life cycle impact assessment results, and other ancillary data or results estimated beyond the LCA (e.g., material health or social impacts). EPDs also include general descriptive data about a product, references to the PCR and program operator, reviewer information, dates of publication and period of validity, qualifying statements about accuracy limitations and lack of comparability with EPDs from other programs, and other explanatory material.

6.8.4 An EPD is Developed by Following a "Product Category Rule"

According to ISO 14025, Product Category Rules (PCRs) are a set of specific rules, requirements and guidelines for developing Type III Environmental Product Declarations (EPDs) for one or more product categories, or a group of products that can fulfill an equivalent function, thus enabling fair product comparisons. While hundreds of LCA studies have been published in peer-reviewed journals, it remains clear that quantitative information can be easily discredited in the absence of clear and detailed guidance on how this information is generated and presented (FTC 2010). PCRs provide guidance on those areas of the LCA standards that are more nebulous and tend to change depending on the goal and scope of the study. This guidance includes: the determination of the functional unit, allocation rules, recommended data sources, impact assessment methods, and additional relevant environmental information that should be included to address environmental concerns (e.g., toxicity in use phase). PCRs should make these calculation rules as clear as possible so that impacts are quantified consistently across multiple studies. ISO 14025 also includes guidelines for the development of PCRs including requirements for the review and verification of the process.

6.8.5 An EPD can contain other Relevant Information beyond the LCA

EPDs can also include information that is not strictly specified in ISO 14025, but referred to as "other relevant environmental information." This category of information is an opportunity for alignment between LCA driven indicators and other indicators that have been insufficiently treated by LCA methods in the past, including water use and scarcity, biodiversity, land use, persistence in environment, individual human toxicity, etc. Many of these non-LCA indicators are of critical importance to the NGO community. Type III LCA-driven product screenings might be criticized for focusing on indirect threats rather than direct threats to high conservation value areas or toxicity susceptibility of vulnerable populations. This category of information provides an opportunity for conversation and possible alignment on indicators which LCA cannot quantify sufficiently, but which are critical to capture true life cycle thinking – and promote the design of healthy, sustainable systems.

6.8.6 Further Information on EPDs and PCRs

EPD's should, by design, provide accurate quantification of environmental attributes of products, communication of these attributes in a standardized and transparent fashion, and permit comparison of one product to another with an EPD in the same product category. As a result the rules (PCRs) behind them need to be very thorough and sound. The level of detail in the PCR depends in large part upon the extent of guidance given by a general guidance document or a standard. General guidance may come from the program operator or other standards body and apply to all products or a large sector of products. If detailed guidance is provided for all products as in some carbon footprinting standards (e.g., the PAS 2050, BSI 2011), less detail is required in the PCR.

The primary intent of PCRs is to guide the development of declarations or labels for products that are comparable to others within a product category. Without this guidance document, there are too many possible permutations in the quantification of impacts to guarantee comparable environmental information (Christiansen *et al* 2006). But because ISO 14025 does not specify a universal program operator, numerous different programs have emerged that are creating PCRs that in some cases duplicate efforts from another program, and create conflicting PCRs (Subramanian *et al* 2011). If duplicate PCRs exist for the same product category, claims made based on different PCRs are not strictly comparable. If PCRs for the same category from different program operators are "aligned" or made consistent, the declarations that originate from them can be made comparable.

References

Baumann H and Tillman A-M (2004) *The Hitch Hiker's Guide to LCA* Studentlitteratur, Lund, Sweden; ISBN 91-44-02364-2.

Boufateh I, Perwuelz A, and Rabenasolo B (2011) Multiple Criteria Decision-Making for Environmental Impacts Optimization *Int J Business Performance and Supply Chain Modeling* 3(1):28-42.

Canis L, Linkov I and Seager T (2010) Application of Stochastic Multiattribute Analysis to Assessment of Single-Wall Carbon Nanotube Synthesis Processes, *Environ Sci & Technol* 44(22):9704-8711.

Christiansen K, Wesnaes M, and Weideman BP (2006) Consumer Demands on Type III Environmental Declarations, prepared for ANEC, Brussels, Belgium.

Cooper JS and Kahn E (2012) Commentary on Issues in Data Quality Analysis in Life Cycle Assessment *Int J Life Cycle Assess* 17:499–503.

DOE (2013) Life-Cycle Assessment of Energy and Environmental Impacts of LED Lighting Products, US Department of Energy, Energy Efficiency & Renewable Energy, April 2013; ssl.energy.gov/tech_reports. html.

EPA (1994) Use of Monte Carlo Simulation in Risk Assessments, EPA/903/F-94/001, US Environmental Protection Agency Region III Technical Guidance Manual, Washington DC, USA.

EPA (1995) Guidelines for Assessing the Quality of Life Cycle Inventory Analysis, EPA/530/R-95/010, US Environmental Protection Agency, Washington DC, USA.

Finkbeiner M (2013) From the 40s to the 70s—the Future of LCA in the ISO 14000 Family (editorial) *Int J Life Cycle Assess* 18:1–4.

FTC (2010) Guides for the Use of Environmental Marketing Claims: Proposed Rule, Federal Trade Commission, Washington, DC, USA.

Grahl B and Schmincke E (2012) "Critical Review" and "Verification" Cannot be used Synonymously - A Plea for a Differentiated and Precise use of the Terms. In Proceedings of the Life Cycle Management (LCM) Conference 2011, Berlin, Germany.

Heijungs R (2002) The Use of Matrix Perturbation Theory for Addressing Sensitivity and Uncertainty Issues in LCA, P2 Infohouse (infohouse.p2ric.org/ref/41/40536.pdf, accessed March 20, 2015).

Heijungs R and Suh S (2002) The Computational Structure of Life Cycle Assessment, Kluwer Academic Publishers; ISBN 978-1-4020-0672-2; 243pp.

Hertwich EG and Hammit JK (2001) A Decision Analysis Framework for Impact Assessment Part 1: LCA and Decision Analysis *Int Journal of Life Cycle Assess* 6(1):5-12.

Ingwersen WW and Stevenson MJ (2012) Can We Compare the Environmental Performance of this Product to that One? An Update on the Development of Product Category Rules and Future Challenges toward Alignment, *J of Clean Prod* 24:102-108.

ISO (2000) ISO/TR 14049:2000(E)Environmental management — Life Cycle Assessment — Examples of Application of ISO 14041 to Goal and Scope Definition and Inventory Analysis, International Standards Organization, Geneva, Switzerland.

ISO (2006a) ISO 14040 Environmental Management - Life Cycle Assessment - Principles and Framework, International Standard, International Organization for Standardization, Geneva, Switzerland.

ISO (2006b) ISO 14044 Environmental management — Life cycle assessment — Requirements and guidelines. International Organization for Standardization, Genève, Switzerland.

ISO (2006c) ISO 14025 Environmental labels and declarations - Type III environmental declarations - Principles and procedures, Genève, Switzerland.

JRC (2010) ILCD Handbook: General Guide for Life Cycle Assessment - Detailed Guide, European Commission Joint Research Center, EUR 24708 EN; 417pp.

Klöpffer W (1997) Peer (Expert) Review in LCA According to SETAC and ISO 14040 *Int J Life Cycle Assess* 2(4):183–184

Rogers K and Seager TP (2009) Environmental Decision-Making Using Life Cycle Impact Assessment and Stochastic Multiattribute Decision Analysis: A Case Study on Alternative Transportation Fuels *Environ Science & Technol* 43(6):1718–1723.

Rowley HV and Peters G (2009) Multi-Criteria Methods for the Aggregation of life cycle impacts, in Proceedings of Sixth Australian Conference on Life Cycle Assessment, Australian Life Cycle Assessment Society, Australia.

Subramanian V, Ingwersen W, Hensler C, and Collie H (2011) Global Comparison of PCRs from Different Programs: Learned Outcomes Towards Global Alignment in a special session on PCR Alignment at the LCA XI Conference, Chicago, Illinois, USA.

Weidema BP (1997) Guidelines for Critical Review of Product LCA. Originally published by the Society for the Promotion of Lifecycle Development (www.spold.org), Brussels. http://lca-net.com/publications/show/guidelines-critical-review-life-cycleassessments/

Weidema B. and Wesnaes M (1996) Data Quality for Management for Life Cycle Inventories - An Example of Using Data Quality Indicators *J of Cleaner Prod* 4:167-174.

Chapter 6 Exercises

1. Connecting Interpretation with the Goal Statement
   Interpreting the results is intended to address the question posed by the goal statement: "the phase of life cycle assessment in which the findings of either the inventory analysis or the impact assessment, or both, are evaluated in relation to the defined goal and scope in order to reach conclusions and recommendations." But sometimes, studies produce surprising results that are not in line with the defined goal. Or, data may not be available to model all the impact categories defined in the goal. Describe how these situations may be approached by an LCA practitioner in order to realign the conclusions stated in the final report after interpretation with the study goal.
2. Uncertainty Analysis
   Uncertainty analysis deals with the effect of imprecise data, i.e. the "noise," related to the results.
   a) Considering the ranges of uncertainty shown as error bars in Figure 6.3, what can be said about the meaningful differences between the products shown? Where do the uncertainty bars overlap and prevent being able to state when one product is less impactful than another?
   b) Looking specifically at the eutrophication (EUT) impact category in Figure 6.3, what factors could cause such a large error bar for electric vehicles (EV)?
3. Sensitivity Analysis
   In a sensitivity analysis, the practitioner assesses all the modeling choices and assumptions in order to determine their impact during interpretation of results. "One-way sensitivity analysis" determines the amount an individual input parameter value needs to change, all other parameters held constants, in order for output parameter values to change by a certain percentage. In the example given in the chapter, changing 498 to 499, a 0.2% change in an input, resulted in a 100% change in the inventory. Explain how the system described in the example does not act in a linear way and how a small change in a parameter can induce a large change in the final results.
4. Contribution Analysis
   a) Contribution analysis is the quantitative determination of the importance of an input in relation to the total inputs to a unit process. Why is this important to know?
   b) Before deciding to exclude materials or processes from the study, it is important to carefully consider the potential effect on study results. Mass contribution is usually the criterion used to identify components for possible exclusion, but, provide an example of a material with a small mass contribution that may have significant impacts on energy or impacts indicator results.
5. Environmental Product Declarations (EPDs) and Other Relevant Environmental Information

Appendix 3 of Chapter 22 (Environmental Product Claims and Life Cycle Assessment) in the LCA Handbook provides additional information for the "Environmental Product Declaration for High-Quality Pasteurized Milk Packaged in PET Bottles." Why do you think the commissioner of the study felt it was important to relay this additional information?

A.3.3.4—Additional Information

A.3.3.4.1 Representativeness of the Farms

Granarolo acquires the raw material from a large number of farms, all situated in Italy, which are divided into classes not only on the basis of the number of animals but also in terms of the quantity of milk produced per day (expressed in litres/day).

The distribution of milk producers according to the distinction made by Granarolo is illustrated in Figure 5, from which it emerges that most of the milk (about 66%) comes from the cowsheds with a daily production of over 3,000 litres/day.

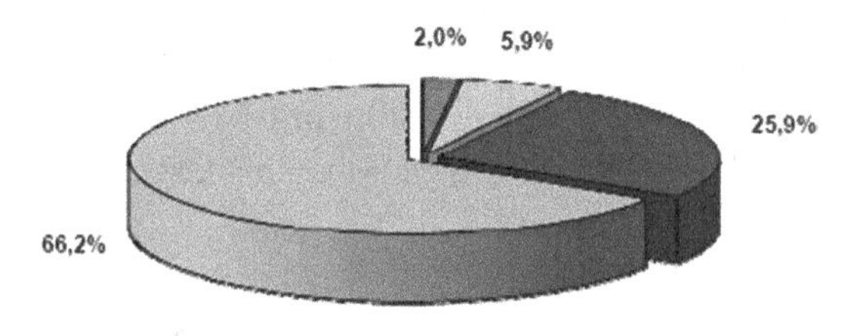

A.3.3.4.2 Primary Milk Production Data

The primary milk production data used for the LCA and to draw up this Environmental Product Declaration refers to a sample of 15 farms; in quantitative terms, the production of these companies was equivalent to 16% of the High Quality milk produced by Granarolo in 2010.

As regards their production, the sample cowsheds belong to the three main categories shown in Figure 5 (501-1000 l/g, 1001-3000 l/g and > 3000 l/g), so as to guarantee that 98% of the High Quality milk produced and delivered to Granarolo is represented (the cowsheds with a productivity of less than 500 litres/day, which account for 2% of the milk were left out of the sample).

A.3.3.4.3 Primary Milk Processing Data

The primary milk processing and packaging data refers to all 5 farms indicated previously and accounts for 100% of the production of Granarolo High Quality milk packaged in one-litre PET containers.

A.3.3.4.4 Use and End of Life of the Bottle

The impacts associated with the use of the milk and management of the primary packaging after use is closely correlated with the consumer behavior.

A.3.3.4.5 Consumption of High Quality Milk

As regards the use phase, the main environmental impact is associated with the storage of the fresh product in the refrigerator considering that the High Quality milk has a life of 6 days following pasteurization.

A.3.3.4.6 End of Life of the Primary Packaging

The white 1-litre PET bottle is a modern remake of the glass bottle (used in Italy until the seventies); the new container was designed with a view to combining the value of tradition with respect for the environment and the need for a more practical container.

In fact, this container is highly practical, light and ensures that the product remains fresh; the PET bottle is not just unbreakable and lighter than the glass bottle, even after it has been opened it can be set horizontally in the refrigerator as it has a liquid-proof cap.

Over the past few years, in order to reduce the consumption of raw materials, Granarolo has gradually reduced the weight of the bottle, by undertaking a project to this aim in 2001, starting from a PET bottle weight of 29 g in 2007, reaching 25 g in 2008 and 23.5 g in 2009-2010, these bottles maintaining their shape and technical characteristics at the same time.

The information given in this document refers to all the phases that can be controlled directly by Granarolo so the end of life management of the package (waste disposal) lies outside the limits of the system assessed in the LCA analysis and the EPD, as indicated in the reference PCR.

Nevertheless, in relation to the handling of PET bottles at the end of their life, it should be pointed out that the environmental impacts depend mainly on the behavior of the end user and the local availability of efficient separate waste collection services; according to the statistics, on average, PET waste in Italy is disposed of as follows:

- Recycling: 41%;
- Waste to energy systems 30%.
- Delivery to dump: 29%;

There are two ways of recycling PET: it can be transformed into secondary raw material or it can be converted into energy as illustrated in Figure 6.

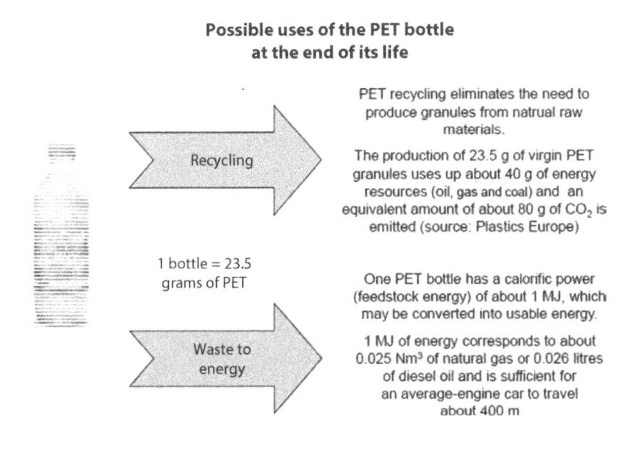

Possible uses of the PET bottle
at the end of its life
Recycling
PET recycling eliminates the need to produce granules from natrual raw materials.
The production of 23.5 g of virgin PET granules uses up about 40 g of energy resources (oil, gas and coal) and an equivalent amount of about 80 g of CO2 is emitted (source: Plastics Europe)
1 bottle = 23.5 grams of PET
Waste to energy
One PET bottle has a calorific power (feedstock energy) of about 1 MJ, which may be converted into usable energy.
1 MJ of energy corresponds to about 0.025 Nm3 of natural gas or 0.026 litres of diesel oil and is sufficient for an average-engine car to travel about 400 m

7

Life Cycle Sustainability Assessment

Abstract
Life Cycle Sustainability Asessment (LCSA) is a proposed approach which integrates the information developed by Life Cycle Assessment (LCA), Life Cycle Costing (LCC) and Social Life Cycle Assessment (SLCA). This has the effect of deepening the scope of mechanisms by including physical, social, economic, cultural, institutional and political considerations in the decision making process. Moreover, a framework based on sustainability enables decision makers to broaden the scope of indicators and shift the object of the analysis from the product system level to the industry sector or whole market level. This chapter describes a possible LCSA framework, analyzing the two components of deepening and broadening, and indicates the future direction of LCSA.

References from the LCA Handbook

Aims of the Chapter

1. Describe how the information generated in conducting an LCA can be merged with other tools, specifically Social LCA and life cycle costing, in a sustainability framework.
2. Outline the steps involved in conducting Social LCA
3. Help readers begin to explore the concept of weak versus strong sustainability.

7.1 Introduction

The concept of sustainability and its assessment is complex and highly controversial, at both the scientific and cultural level. Different perspectives and schools of thoughts have been developed, leading to the debate between weak and strong sustainability[1], with the question of substitutability at the core. Moreover, each discipline has framed its own interpretation of what sustainability entails.

Without discussing the appropriateness of defining sustainability as a science, we have to recognize that a scientific analysis is necessary to answer questions regarding the sustainability of projects, technologies, policies, and so forth. How to demonstrate *what* might be sustainable and *how* to measure it (Graedel and Klee 2002) require a scientific approach, which is also open to interpretation.

Many different approaches and metrics have been developed over the past decades: tools for environmental assessments, tools for economic modeling and assessments, approaches for sociological analysis, approaches for integrated assessments, methods and tools for futures studies, and participatory approaches. Several sustainability indicators have been developed by the UN, the OECD and the EU, as well as by companies and NGOs, often subdivided into groups covering the economic, environmental and social dimension. These indicators are an important ingredient in the process of communication, benchmarking and decision-making and for this reason their scientific validity is a crucial factor. Any company can claim its products are sustainable, and any NGO can deny this, but only scientifically based analysis and methods can provide a rational basis for decisions and arguments.

Despite the great number of methods and approaches developed, the broader notion of sustainability is still not sufficiently reflected in the existing assessment frameworks and tools. In fact, each of them addresses only specific aspects and thus fails in describing the complexity of mechanisms and linkages, which are inherent to any sustainability evaluation. An interdisciplinary integration is required, defined as *the activity of critically evaluating and creatively combining ideas and knowledge to form a new whole or cognitive advancement. It contributes to solving complex problem by providing a systematic approach to combining and interrelating insights grounded in commonalities while taking into account differences* (Bruins *et al* 2009).

The concept of integration echoes that of consilience, i.e. unity of knowledge, which indeed provides a nice and effective representation of what sustainability entails. The idea behind is that each branch of knowledge studies a subset of reality that depends on factors studied in other branches. In the context of sustainability analysis, this means that different disciplines are necessary and need to converse with each other: empirical knowledge on the one side, with physical, environmental, economic, technical modes; and normative positions on the other side, with ethical and societal values. Moreover, building upon the concept of linkages pointed out by Graedel and van der Voet (2010) and broadening it, these domains are inter- and intra-connected, and the understanding and the identification of these linkages requires investigation.

---

[1] This implies a range of sustainability activity. In layman's terms, weak sustainability refers to activities that allow for the continued depletion or degradation of natural resources, but at a slower or impeded pace. Strong sustainability, at the extreme end, does not allow for the use of non-renewable resources.

7.2 Life Cycle Assessment and Sustainability

Being a global concept that covers present and future generations, sustainability inevitably calls for a system-wide analysis. Such a system perspective is at the core of the life cycle approach, which can provide a valuable support in sustainability evaluations. In fact, the main argument for a life cycle view is to prevent false optimization and wrong choices, like the burden shifting within or between each pillar (environment, economy and society) or to the future.

Life Cycle Assessment (LCA), as the most mature life-cycle based method, is a good candidate for facing the challenges posed by sustainability questions. In this regard, the environmental policies at European level provide a valuable example of life cycle thinking (LCT). In fact, LCT and LCA are central themes of the recent Sustainable Consumption and Production (SCP) Action Plan (COM 2008), as they are in the Eco-Design Directive (EC 2009), the Waste Framework Directive (EC 2006), the Thematic Strategy on the Sustainable Use of Natural Resources (COM 2005), and the Environmental Technologies Action Plan (COM 2004). Moreover, in the Integrated Product Policy Communication, the European Commission states that "LCAs provide the best framework for assessing the potential environmental impacts of products currently available" (COM 2003). The recently published International Life Cycle Data System (ILCD) Handbook (JRC-IES 2010a), made available through the European Platform on LCA, is a further confirmation of the importance of LCA as a decision-supporting tool in contexts ranging from product development to EU policy making. In fact, the Handbook[2], a series of technical guidance documents to the ISO 14040-44 standards (ISO 2006a,b), is aimed at serving as a basis for comparable and reliable LCA applications in business and public decision-making. However, LCA has been developed and standardized firstly for evaluating the environmental potential impacts of goods and services. It simply applies a linear static model based on technological and environmental relations in inventory and impact assessment phases respectively and it is moreover restricted to impacts on the environment (Heijungs *et al* 2010). While the "simple" LCA model is also due to the useful intention of keeping it operational and limiting its complexity and inherent uncertainty, these features are in many ways in contradiction with the requirements of a sustainability evaluation from the epistemology point of view. In fact, sustainability clearly shows distinctive marks of complexity theory, which presently cannot be dealt with in LCA: non-linear relationships, feedback loops, emergent phenomena, and tangled connections among the parts.

The biofuel example clearly demonstrates (Zamagni *et al* 2009) the shortcomings in the current way of applying LCA for sustainability evaluations. In particular with reference to the environmental pillar, the seminal study by Searchinger *et al* (2008) found that most of the previous LCA studies provided only a limited answer to the problem because, by excluding emissions from land use change, they failed in accounting for the indirect effects, i.e. those taking place outside the biofuel value chain. Also the effects on global food prices and the induced clearing of rain forests, for example, have not been included in the analysis, missing major social, economic and environmental

[2] It includes explicit and goal-specific methodological recommendations, a multi-language terminology, a nomenclature, a detailed verification/review frame and further supporting documents and tools.

Table 7.1 Dimensions of Questions for Sustainability (adapted from Heijungs *et al* 2007; Guinée *et al* 2009).

Dimensions	**Main characteristic(s)**	**Explanation/example**
Time horizon	Short term/long term	
Perspective	Past, present, future	
	Diachronically/ synchronically	Before-after comparisons vs comparisons in the same time frame
Purpose	Satisfying and optimizing	
	Optimizing or choosing	
	From occasional choices to more broadly framed questions	Is it better from a sustainability point of view to take the plane or the car today, knowing that the flight will take place anyway and will fly half empty and I will drive with three people in my car today? Or Is it better from a sustainability point of view to go by air or by car, under some general assumptions about average conditions?
Level of aggregation	Micro	Choice of biomass for car fuel
	Meso	Second generation biomaterial and energy technologies
	Macro	Bio-based economy

consequences. That of biofuel is just one example, but questions for sustainability may be many, may include different dimensions, as shown in Table 7.1, and they cannot always find an all-round answer in LCA for the limits discussed above.

These dimensions are not sharply defined but are usually combined. In fact, the time horizon and the perspective are defined for the different types of purposes, which in turn can be defined at different levels of aggregation. There is also a lot of discussion about the exact distinction among micro, meso and macro levels, since many types of questions fall in between. Difficulties arise because the levels are intertwined. In fact, even if the outcome may be that the micro level questions are essential in the end, the meso and macro level questions may bring focus to relevance of the micro level, and may help shape micro level questions. Moreover, there is the question of emergent phenomena and properties, according to which the system at higher level (i.e. macro) shows characteristics and behavior we could not understand from the observation and analysis of its main parts, at micro level. This usually happens because there are causal relations or mechanisms which act at different scales, with feedback in the system.

How to link these different levels, i.e. how many decisions at the micro level work out at the macro level, for total society and *vice versa* (i.e. backcasting, how to cascade global objectives down to individual actions and what will be the effect of each of the steps between (Guinée and Heijungs 2011) is the problem at the core of sustainability analysis which requires further sophisticated and broader model than LCA.

Attempts to make LCA practice more suitable for sustainability evaluations have been ongoing for some time. Andersson *et al* (1998), for example, examined the feasibility of incorporating the concept of sustainability principles in each phase of LCA. Four socio-ecological principles were identified:

- Substances from the lithosphere must not systematically accumulate in the ecosphere (i.e. the use of fossil fuels and mining must be radically decreased);
- Society-produced substances must not systematically accumulate in the ecosphere;
- The physical conditions for production and diversity within the ecosphere must not systematically deteriorate (i.e. more efficient and careful use of areas productive for agriculture, forestry and fishing);
- The use of resources must be efficient and meet human needs

Later, in the year 2000, The Natural Step approach was proposed (Upham 2000) with principles for sustainability to inform LCA impact categories and using backcasting in the LCA framework. These approaches highlighted the need of a perspective wider than that of LCA, as proposed later by Hunkeler and Rebitzer (2005) and formalized with the Life Cycle Sustainability Assessment framework (Klöpffer 2008), according to which:

$$LCSA = LCA + LCC + SLCA,$$

where LCC and SLCA are Life Cycle Costing and Social Life Cycle Assessment, respectively.

The three methods, which have a different degree of development, are applied at product-level, independently one from another, under specific consistency requirements but without considering the mutual relations which can arise.

Keeping the life cycle framework as a reference, the need exists to explore whether LCA is suitable for evaluating sustainability and what it is necessary in terms of research efforts. More in detail, following the principles of transdisciplinary integration discussed before, researchers should investigate what is available in other life cycle methods and models already that might be further develop; what is not yet there but is developing in other domain of sustainability analysis and what is available in non-life cycle models but might be incorporated in a life cycle framework for sustainability. A framework for sustainability evaluations should be able to take into account broader externalities, broader interrelations and different application/user needs, to deal with the forecasting of system behavior and technology evolution, normative choices, uncertainties and risks. Supplying this information in a consistent way at all levels of society is a central challenge for research for sustainability.

7.3 A Framework for LCSA

Addressing sustainability as discussed above requires going beyond LCA as standardized in ISO, in order to overcome those limits which presently make LCA applicable to a wide range of products/systems/technologies. Therefore, in order to develop a life-cycle based analysis for sustainability we need to move into two main directions:

- Deepening the scope of mechanisms and/or a particular mechanism. Deepening can be achieved by going beyond the focus on technological and environmental mechanisms and including also physical, social, economic, cultural, institutional and political ones (Heijungs *et al.* 2010). On the other side, deepening means also further sophisticating the modeling, for example adopting spatially differentiated models;
- Broadening the scope of indicators and/or the object of the analysis. Broadening can be achieved by staying in the realm of environmental indicators, by extending them in numbers, or by going beyond the focus on environmental aspects and including also the economic and social ones in the analysis. Another example of broadening is the shift of the analysis from individual product systems to sectors, baskets of commodities, markets or whole economies.

A proposed framework for LCSA is depicted in Figure 7.1 (Guinée *et al.* 2011). It introduces the following characteristics:

- It focuses on mechanisms, i.e. causal relations that connect two activities. The following are considered in the framework: technological relations; environmental mechanisms; physical relations; economic relations; social, cultural and political relations. Overall they constitute empirical knowledge. Also, normative positions are included in terms of ethical positions and societal values (Figure 7.2). Adding mechanisms means deepening the analysis. This can be done in each box of Figure 7.1, either a techno-system box or an environmental, economic or social performance box.
- Inventory analysis and impact assessment are merged into one modeling phase. Both empirical knowledge and normative positions are considered and, thus, a clear separation among behavior, technology, environmental processes, etc. becomes difficult and artificial. For example, the human toxic impact due to PM10 releases from cars depends, besides the characteristics of the technological system, on human behavior (how much time people spend outside, whether air quality control equipment are installed in offices or houses) and public policy (e.g. the distance between highways and residential areas). This simple example shows that from the inventory analysis to the impact assessment there is a complex pathway of causes, effects and feedback, which no longer cannot be dealt with separately.
- It broadens the object of the analysis to meso and economy-wide levels.
- It broadens the set of indicators so to include also economic and social performances.

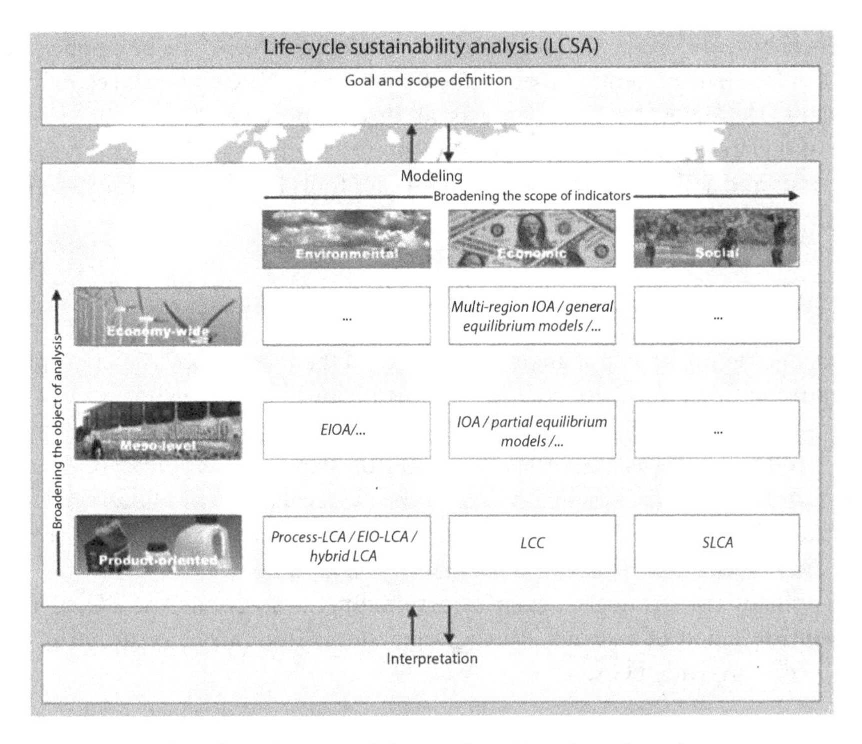

Figure 7.1 Framework for Life Cycle Sustainability Analysis (Guinée *et al* 2011).

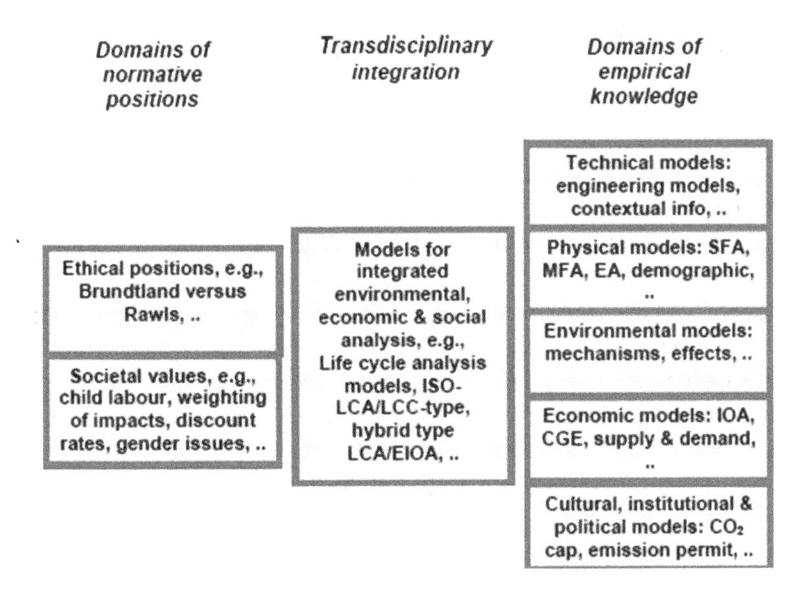

Figure 7.2 Empirical knowledge and normative positions in the framework at the basis of the transdisciplinary integration (Zamagni *et al* 2009).

For each box, different methods and models exist which can be applied, among which LCA for environmental aspects, Life Cycle Costing (LCC) for the economic ones and Social LCA (SLCA) for social aspects, all three at product level.

Most of the present developments in LCA fit into the framework, as described in the next sections and a plethora of other methods and tools have been already identified as potentially useful for working with the framework.

7.3.1 Broadening of the Object of Analysis

With broadening the object of analysis, we refer to the fact that LCA can be applied to a wide range of things (Guinée *et al* 2011). This includes the following:

- products, such as a beverage container or diapers. This corresponds to a product-level LCA, which is in fact the classical target of LCA;
- larger system, such as the energy provision of a region or a waste treatment facility. This corresponds to a sector-level LCA,
- policies and strategies, such as a recycling strategy for the EU or the introduction of an emission trading system. This can be regarded as an economy-wide LCA.

Obviously, the level of products, sectors and the full economy are not clearly demarcated categories and several applications fall in between. Equally obvious is that stretching LCA to address broader objects requires a reassessment of purpose, principles and methods.

In fact, many of the original model assumptions of mainstream LCA are justified in the perspective of an object of limited realm, i.e. of a product-level LCA. This includes, e.g., the ceteris paribus assumption, the concept of a functional unit, system boundary and allocation, and the linear and marginal calculus in inventory and characterization (Guinée *et al* 2002). In LCA studies at the sector-level, some of these assumptions no longer apply. For instance, there is often not a simple functional unit. Waste treatment systems treat a mix of waste products, and the exact composition scenario determines how the processes work in terms of input requirements and emissions, and which secondary products (electricity, metals, road construction materials, etc.) can be expected. This explains that the choice of functional unit and allocation is much more debated and crucial than for the product-level LCA. For the economy-wide LCA studies, even more issues pop up. These include, for instance, an unacceptably large deviation from the *ceteris paribus*, linearity and marginality assumptions of process data and impact models.

In developing LCSA, we should well understand the assumptions of the LCA model in relation to the object of analysis, and be prepared to use a more complicated approach or model for questions that necessitate so. This may involve stepping outside the accepted framework in particular situations. For instance, in comparing different food patterns, the classical physical mechanisms in the inventory model should be enhanced by economic mechanisms (new prices, budget shifts) in order to get a more realistic result (Tukker 2011).

Altogether, we see that broadening the object of analysis is connected to deepening mechanisms. But deepening mechanisms as such is a technical phenomenon, while broadening the object of analysis relates to the questions posed.

7.3.2 Broadening of the Spectrum of Indicators

While broadening the object of analysis is related to the question, broadening the spectrum of indicators relates to the results. The first LCA studies addressed energy and waste, but pretty soon, other environmental aspects were included, although often in slightly primitive ways. From the idea of the critical volume for pollutants, gradually there emerged the idea of adapting models for risk analysis, including fate, exposure, effect, and – more recently – damage. Nowadays, mainstream LCA (JRC-IES 2010b) addresses a dozen of impact categories at the midpoint level (including climate change, human toxicity, acidification and resource depletion) or three wider areas of protection at the endpoint level (such as human health, ecosystem quality, and resource availability). However, in some cases this is not enough, and especially in the context of LCSA we need more.

7.3.2.1 Broadening by Inclusion of Additional Environmental Indicators

The idea of impact indicators is always that some issue is addressed by means of a proxy issue. We address infrared radiation in midpoint LCIA, not because infrared radiation is in itself interesting, but because we believe it is correlated to climate change. However, even climate change as such is not of ultimate interest, since the causal chain continues to a plethora of disturbing issues, such as sea level rise, change in agricultural productivity, change in vegetation, etc. Many of these issues are difficult to model, but some of them may still be expected to correlate to some extent with the previous variable. For others, this correlation may be much weaker, or unknown. It may thus be appropriate to introduce some of these as separate impact categories.

Besides this, there are many environmental issues that are not often addressed in LCA. The inventory data are unknown, the impact models are lacking or not agreed upon, or LCA practitioners (or database or software developers) are simply not aware of these issues. Researchers have been addressing these "non-traditional" impact categories for decades, and some of these have gained acceptance, but others are still in the phase of development. Examples of such impact categories are noise, accidents, desiccation, erosion and salination (JRC-IES 2010b), but there are more (biodiversity, effects of GMOs, landscape fragmentation, electromagnetic waves, light pollution). For some categories, like ionizing radiation, accepted practices have been defined (JRC-IES 2010b), but most LCA studies still ignore them. For other categories (like thermal pollution (Verones *et al* 2010) and ecosystem services (Zhang *et al* 2010), interesting approaches have been developed but so far lack general acceptance and/or implementation.

It is clear that in a sustainability perspective, not only the classical impact categories can be of interest. Depending on the topic, other impact categories can be relevant. And as LCSA will allow for broader questions and deeper mechanisms, unexpected impacts may become crucial. In a classical LCA on fossil fuels versus biofuels, climate change and resource depletion may be decisive. In an LCSA that does not look on the functional unit of 1 MJ of energy, but on an economy-wide application and that includes trade-offs, indirect land use suddenly becomes a critical aspect.

7.3.2.2 Broadening by Inclusion of Economic Indicators

With the broadening of sustainability into sustainable development came the broadening of environmental awareness into the triple bottomline: People, Planet, Profit

(Elkington 1998). It emphasizes that a sustainable society provides food, housing, clothes, etc.; is socially sound in terms of employment, equity, democracy, etc.; and has good environmental conditions and abundant nature, in terms of clean air, wildlife, forests, etc.

As such, the broadening of LCA into LCSA should address more than environmental issues; it should also address economic and social issues (Hunkeler and Rebitzer 2005). These two categories of impacts are sometimes difficult to distinguish, so in literature you may find that what one study labels as an economic indicator will be labeled as a social indicator by a second study, and as a socio-economic indicator by a third one. In practice, the economic dimension in LCA is most often restricted to the business point of view, using the concept of life-cycle costs (LCC). Other economic indicators are often brought under the social dimension, such as employment.

LCC "summarizes all costs associated with the life cycle of a product that are directly covered by 1 or more of the actors in that life cycle" (Hunkeler *et al* 2008). It is, in fact, a complete LCA procedure, including a goal-and-scope-like phase, an inventory-like phase, and an impact assessment-like phase. In principle, LCC could deliver life cycle-based economic information that could fit into an LCSA. Unfortunately, the LCC framework and the LCA framework differ, not only in terminology, but also in content (Huppes *et al* 2004). Important deviations occur on topics like system boundary and dealing with time.

Besides LCC, the economic dimension can be provided by various techniques and methods. For example Total Cost Accounting (TCA) and Total Cost of Ownership (TCO) have been/could be used in combination with LCA. The differences among them rely mainly in the concept of cost (private versus social), but other parameters like time path specification and discounting need to be taken into account. LCC and TCA are considered as the economic counterpart of LCA while TCO as a specific case of LCC, where the assessment takes the perspective of the product user/consumer (Schepelmann *et al* 2008). As far as externalities are concerned, the ExternE methodology has been developed to get a full assessment of external costs of environmental burdens. Making use of LCA, in combination with Impact Pathway Analysis and Cost Benefit Analysis (CBA), ExternE has been proposed as applicable to the whole scale of LCSA, i.e. to a technology, a sector or to the whole economy (Friedrich 2011). Regarding CBA, used for assessing the total costs and benefits of a project or an activity, it shows similarity with LCA in the way in which biophysical and social externalities are accounted for (Weidema 2011). However, also many differences exist, which make the combination not an easy task.

7.3.2.3 Broadening by Inclusion of Social Indicators

Besides the cost aspect, also socio-economic issues are important for a broader sustainability assessment. Social impacts have been largely discussed in the scientific literature and different perspectives emerged. For example, Assefa and Frostell (2007) have defined social impacts as those related to individual well-being and to the interactions among individuals, while Vanclay (2009) more generally as changes occurred in the way people satisfy their needs as citizens embedded in the society (way of life, culture, community, political systems, health and well-being, fears and aspirations, etc.).

In the LCA community a common perspective is that social impacts are those that may affect stakeholders along the life cycle of a product and may be linked to company behavior, to socio-economic processes and to impacts on social capital. In the last few years, social life cycle assessment (SLCA) has been receiving more and more interest (Dreyer *et al* 2006; Weidema 2006; Ciroth and Franze 2011; Franze and Ciroth 2011). Recently, UNEP-SETAC felt time to condense the state of the art into guidelines (Benoît and Mazijn 2009) and to make available methodological sheets which provide also examples of indicators (site-specific and generic) that can be used in the assessment. The backbone of the SLCA framework is its stakeholder perspective: social and socio-economic impacts are observed in relation to five main stakeholder categories (workers, local community, society, consumers, value chain actors), which are linked to specific impact categories, and whose relevance is determined by the object and goal of the analysis. SLCA is presented in greater detail in section 7.4.1.

Moving from the product-level to the sector and economy-wide ones, a method like the Social Impact Assessment (SIA) becomes relevant. The concept of social acceptance, which is at the core, was further elaborated and led more recently to the idea of social compatibility[3], used mainly in the context of technology assessment. The application of the principles of SIA to the life cycle framework should be further explored as they offer a first opportunity for broadening the level of the analysis.

7.3.2.4 Combining/Integrating Dimensions

In classic LCA, the weighting step that was assumed to pave the ground for an unambiguous assessment, created quite some controversy. By broadening the spectrum of indicators, this problem gets multiplied. On top of a weighting of environmental impacts, LCSA requires weighting of costs and social indicators. This aspect has received very little attention by method developers.

Alternatively, a number of schemes has been introduced to form ratios and/or to use visual means of combining a number of indicators. The most well-known of these is eco-efficiency (EE). EE is a concept around which quite some unclarity has been around, but which to an increasing extent is understood as the ratio between an economic and an environmental indicator, or the other way around (Huppes and Ishikawa 2007). Of course, this presupposes that both types of indicators (economic and environmental) are already expressed as a single indicator. This either requires a prior partial weighting within each sustainability domain, or a rigorous selection, e.g., the carbon footprint as a proxy for overall environmental impact.

In a visual context, BASF's eco-efficiency diagrams (Saling *et al* 2002) have become a famous tool for communicating economic and environmental information at one time. A three-dimensional extension by BASF is the SEEBALANCE (Saling *et al* 2005), combining one economic, one environmental, and on social indicator. More recently, the Life Cycle Sustainability Triangle and the Life Cycle Sustainability Dashboard (Finkbeiner *et al* 2010) have been proposed as a multi-criteria evaluation approach to deal with LCA, LCC and SLCA indicators resulting from the application of the framework of Klöpffer (2008).

[3] According to the idea of social compatibility, decisions made in the economic, political or technological sphere should take into account the societal values and visions

These few examples demonstrate that a key challenge for a LCSA resides in presenting and communicating results. Clarity shall be at the core and oversimplifications should be avoided since they could hamper transparency, which is a *sine qua non* requirement for any assessment method.

7.3.3 Deepening

Deepening primarily refers to an increase of the level of sophistication[4] in modeling intra-system relationships and in modeling impact indicators. But it can also refer to a modeling of mechanisms that were not modeled in classic LCA.

Classic LCA is based on a model that can be characterized as follows:

- the intra-system relationships are primarily of a physical nature;
- the impact models are primarily modeled on the basis of environmental relationships;
- the models are either linear or linearized;
- the models are not dynamic and spatially explicit only to a very limited extent.

As an example, consider an LCA of a can for beverages. Producing the can means that there is a need for aluminum and electricity. This represents a physical necessity. Other physical necessities follow from this by moving further upstream. One of the physical flows from these activities is CO_2. What happens with this substance once emitted, its fate in the atmosphere, its contribution to disturbing the climate system, is an environmental mechanism.

7.3.3.1 Increasing Sophistication in LCI Modeling

LCA has so far concentrated on physical mechanisms in the inventory and on physical and environmental mechanisms in the impact assessment. As a consequence, the modeled relationships are getting better and better. Nevertheless, LCA scientists are expanding the horizon further and there are plenty of examples in literature of interesting approaches which show that sophistication can be achieved in many ways. .

Consequential LCA, dealing with non-linear relationships, hybrid approaches, temporal differentiation, adding optimization strategies through linear and non-linear programming, and the numerous developments occurred in impact assessment, ranging from regionalization to the development of new characterisation methods and adding expert knowledge (such as biotic ligand models for ecotoxicity) are just some examples. Focusing on the LCI phase, Hybrid approaches combining input-output analysis (IOA) and LCA are considered very promising. In fact, with their flexible computational structure, they potentially open up new perspectives towards expanding the scope of LCA applications to higher scales of analysis, from "*micro questions on specific products, to meso questions on life styles up to macro questions in which the societal structure is part*

[4] Increasing the sophistication of the present modeling in LCA is a process aimed at adding more realism and thus, at increasing the fidelity of the model, i.e. its ability to capture the complexity and those interrelations within the system that are really meaningful.

of the analysis" (Heijungs *et al* 2010). We are still in the realm of steady-state models, in which changes in time are ignored. Nevertheless, dynamics are of fundamental importance both in industrial and political contexts, and ignoring them could lead to reduced relevance of some results. Attempts to introduce time dimension in the LCA modeling exist, but the applications are still controversial and open questions exist whether the use of scenarios, especially for decisions related to the long period, could be more relevant and feasible. In fact, by moving from a marginal functional unit to sector and economy-wide questions, scenarios may support the analysis in providing a description of how the world may change.

These new approaches and developments are ongoing, and they still need further improvements to become more applicable and a common practice in LCA.

We should not forget also that developments on the methodological sides should go hand in hand with improvements on tools, i.e. software and databases. In fact, despite the broad choice of LCA tools for different sectors and applications, the capabilities for the analyses discussed above are not fully addressed.

Efforts should be spent on improving the existing tools abut also in designing new ones, making them able to work with different data sources (e.g. economic and environmental), increasing their computational ability (they need to work with a notably increasing quantitative of data) and, mostly, making them flexible and transparent.

7.3.3.2 Economic and Behavioral Mechanisms

More realism cannot only be added to the LCI by deepening the existing physical modeling, but also by adding economic and behavioral mechanisms. These two are treated together here, as the boundary is not always clear.

An example of a relevant mechanism is the rebound effect. This refers to the situation that a more efficient technology is used to such an extent that the intended gain is partly spoilt or even turned into a loss. This may happen, for instance, when fuel-efficient cars induce the user to use it more often, or by energy-efficient light bulbs that can be seen in many gardens.

Another example is that environmental awareness may suggest eating less meat. As meat is often one of the dominant expenditure categories in food, consumers will spend less on food, and may spend their savings on polluting activities like traveling by airplane.

Such mechanisms are getting more important with the broadening of the object of analysis, at least for the LCI. But also in the LCIA, economic and behavioral mechanisms can become important. Impacts of climate change will be mitigated, by building dams, by constructing hurricane-proof buildings, and by using more air conditioners. Including such mechanisms will create a feedback loop from the impact assessment to the inventory, or even to the goal definition (Heijungs *et al* 2009). In literature approaches to deal with rebound effects have recently been proposed (Girod *et al* 2010) and among the methods to quantify them, the own-price elasticity of demand, the cross-price elasticity of demand, and general equilibrium models are those commonly used so far.

7.3.3.3 Deepening LCA and Consequential LCA

Consequential LCA (CLCA) is a clear example of how the deepening can be achieved in LCA. In fact, being CLCA aimed at describing the effect of changes within the life cycle and given that changes lead to a series of consequences through chains of cause-effect relationships (Curran *et al* 2005), mechanisms are at the core of this modeling technique. Present focus is on market mechanisms, i.e. those driven by the interaction of supply, demand, and prices. These mechanisms are dealt with exogenously[5] through the inclusion in the system under study of the affected processes, defined as those that respond to the change in demand driven by the decision at hand.

We consider the case of biofuels as a guiding example. The production of bioethanol represents a new energy product into the market, causing changes in prices and in volume within the energy market but also outside, affecting other commodities. Among the possible consequences, for example it could happen that more corn would be required for bioethanol, squeezing out corn for food and also land use for wheat production. Both prices will increase, with still other products being squeezed out and rising in price. The chain of consequences to analyze could be very long and complex. However, in CLCA simplifications are adopted, for example in relation to the number of markets dealt with simultaneously, to the scale of consequences or to the complexity of substitution mechanisms.

Going beyond and thus not limiting the approach to the market mechanisms, the consequential logic could be extended so to consider CLCA not a modeling principle with defined rules but an approach to deepen LCA, to include more mechanisms. Which ones to include is a tricky question, since they can show up everywhere, involving a variety of domains. Market mechanisms are part of broader economic mechanisms, which are related to concepts like employment and growth. These in turn function within a cultural, social, political and regulatory context. Taking this complex chain of consequences into account would require important developments in CLCA, the main one being the removal of those constraints that presently are considered as fixed entities. Thus, in perspective CLCA offers already the conceptual basis for a life cycle sustainability assessment, proposing itself as a way for dealing with the modeling step in the LCSA framework. However, it is a necessary first step to improve present capabilities of CLCA to fully address market mechanisms, before extending it to a more complex analysis. Efforts are necessary to clarify when and which market information is important, to improve long-term forecasting techniques and to identify the affected processes. In this regard, a great contribution could be given by scenario modeling, since it would provide a more sound-scientific basis to model specific product-related futures for example with respect to technology development and market shifts.

7.4 Social Responsibility

Social responsibility is a term introduced in the 1950's (Bowen 1953). It is used to refer to the responsibilities enterprises can assume in order to contribute to sustainable

---

[5] They are derived from economic models or outlooks in specific sector and then included as input into LCA.

development (UNEP/SETAC 2009, ISO 2010). In order to be socially responsible, it is normally understood that companies should go beyond the law, "recognizing that compliance with law is a fundamental duty of any organization and an essential part of their social responsibility" (ISO 2010). A crucial aspect of an organization supply chain social responsibility is the respect for human and worker rights. Human Rights are defined by the Universal Declaration of Human Rights adopted in 1948 and operationalized by the means of international conventions and treaties. Human Rights protection may be an important component of a country's law or the law may lack requirements. It may also be that a country's legal system lacks enforcement for a number of reasons. The UN Protect, Respect and Remedy framework adopted in 2008 was developed to help clarify duties and responsibilities regarding the observance of human rights in a globalized context, where "widening gaps between the scope and impact of economic forces and actors, and the capacity of societies to manage their adverse consequences, were unsustainable" (Ruggie 2008).

In addition to human rights, ISO 26000 defines five other social core subjects to be managed by organizations: labor practices, organizational governance, fair operating practices, consumer issues, and community involvement and development (ISO 2010). ISO 26000 provides organizations with guidance on social responsibility. Because of its extensive multi-year and multi-stakeholder process under the ISO umbrella, it is considered to be a pillar of social responsibility implementation and management (Capron, Quairel-Lanoizelée and Turcotte (eds) 2010).

7.4.1 The Social LCA Framework

Social LCA (SLCA) guidelines integrate all the ISO 26000 core subjects within a framework to operationalize assessments. The framework defines 31 social subcategories of impact. Subcategories are "socially significant themes or attributes classified according to stakeholder groups and/or impact categories" (Benoît-Norris *et al* 2011). The use of subcategories in SLCA was inspired by its use in Environmental LCA. Indeed, impact categories may be subdivided into subcategories when they are too heterogeneous to allow for scientifically valid aggregation (UNEP/SETAC 2009; Udo de Haes *et al* 1999).

Five types of stakeholder groups classify subcategories: Worker, Local Community, Society, Value Chain Actors, and Consumer. The subcategories may also be classified by impact categories such as Human Rights, Working Conditions, Health and Safety, Cultural Heritage, Governance, and Socio-economic Repercussions. The framework was developed to be in line with ISO 26000 and impact subcategories can also be mapped to ISO Social Responsibility core subjects. The impact categories represent social issues of interest that will be expressed regarding the stakeholders affected (UNEP/SETAC 2009). The main value of using impact categories in assessments not relying on causal chain modeling is to summarize results to stakeholders. The table below maps subcategories to impact categories. Two subcategories not mentioned in the Guidelines were added to the table based on experience with using the framework. Education and training are an important piece of an organization's efforts to improve their social responsibility and the published version of the Guidelines framework lacked a proper subcategory to account for it. It is also very worthwhile to be able to identify whether organizations have implemented Social Responsibility management systems when conducting site-specific studies.

The table also points out which subcategories it only makes sense to use for site-specific studies. All others can be used to offer insights at the sector, country-specific sector or country level in generic or site-specific assessments. A country-specific sector is a sector of the economy located in said country, ex. vegetable and fruit sector, Chile (Benoît et al., 2010).

Location information is a must in SLCA for most subcategories of assessment. It does not need to be precise; it can indicate the country or the region where the production activity takes place but it is necessary - most of the time. However, for some subcategories the technology, process of production, or the industry type may play a greater role than the location. In that case, generic assessment may be conducted without the location information.

Subcategories that may not absolutely need location information to be assessed at the generic level are: Worker, Health and Safety (i.e., in the case of agriculture and pesticide use), Worker, Working Hour (i.e., in the case of apparel, sewing), Consumer, Health and Safety, Local Community Safe and Healthy Living Condition (i.e., in the case of the chemical industry),

When assessing site-specific impacts, some subcategories may also target issues that may not be as sensitive to location information as others may be. For example, some corporate policy lies at the corporate level and may not change according to factory location.

At the site-specific level: Society, Public Commitment to Sustainability Issue, Technology Development, Value Chain Actors, Fair Competition, Value Chain Actors, Promoting Social Responsibility, Value Chain Actors, and Respect of Intellectual Property Rights may be assessed without location information.

The methodological sheets developed as complementary material to the Guidelines provide examples of indicators that might be used for generic and site-specific studies without proposing a one fit all approach (Benoît *et al* 2011). Instead, it proposes to use information on the subcategory in the format and indicator shape that is the most readily available. The next section will present more details on the process of an SLCA study.

7.4.2 Iterative process of Social Life Cycle Assessment

SLCA is an iterative technique that goes through the same phases as a typical E-LCA study would. It usually follows the following sequence: Definition of the Goal and Scope, Life Cycle Inventory Analysis, Life Cycle Impact Assessment and Interpretation. Because of its iterative nature, the assessment will generally go through phases a number of times.

The SLCA Guidelines recommend carrying a social hotspots assessment in order to prioritize data collection activities. Because product systems may include over a thousand unit processes, it is necessary to focus resources. Criteria may be used to select unit processes or a country-specific sector that should be investigated further. In a later section, the social hotspots database (SHDB) system will be presented as a recent development to the field of SLCA.

Table 7.2 Subcategories Mapped to Impact Categories for Use in Social LCA.

Stakeholder Group	**UNEP/SETAC Subcategory**	**Site-Specific only**	**UNEP/SETAC Impact Category (Type 1)**					
			Governance	**Human Rights**	**Working conditions**	**Socio-Economic Repercussions**	**Health and safety**	**Cultural Heritage**
Worker	Freedom of Association and Collective Bargaining				✓			
	Child Labor				✓			
	Fair Salary				✓			
	Working Hours				✓			
	Forced Labor				✓			
	Equal Opportunities/Discrimination				✓			
	Health and Safety						✓	
	Social Benefits/Social Security	SB ✓			✓			
	Education and Training	✓			✓			
	Management system	✓	✓					

(Continued)

Table 7.2 Cont.

Consumer	Health and Safety						✔	
	Feedback Mechanism	✔	✔					
	Consumer Privacy		✔					
	Transparency	✔	✔					
	End of Life Responsibility					✔		
Local Community	Access to Material Resources					✔		
	Access to Immaterial Resources					✔		
	Delocalization and Migration			✔				
	Cultural Heritage							✔
	Safe and Healthy Living Conditions						✔	
	Respect of Indigenous Rights			✔				
	Community Engagement	✔	✔					
	Local Employment	✔				✔		
	Secure Living Conditions			✔				
Society	Public Commitments to Sustainability Issues	✔	✔					
	Contribution to Economic Development					✔		
	Prevention and Mitigation of Armed Conflicts			✔				
	Technology Development					✔		
	Corruption		✔					

Value Chain Actors	Fair Competition	✔	✔					
	Promoting Social Responsibility		✔					
	Supplier Relationships	✔	✔					
	Respect of Intellectual Property Rights		✔					

Table adapted from Benoît and Vickery-Niederman 2010

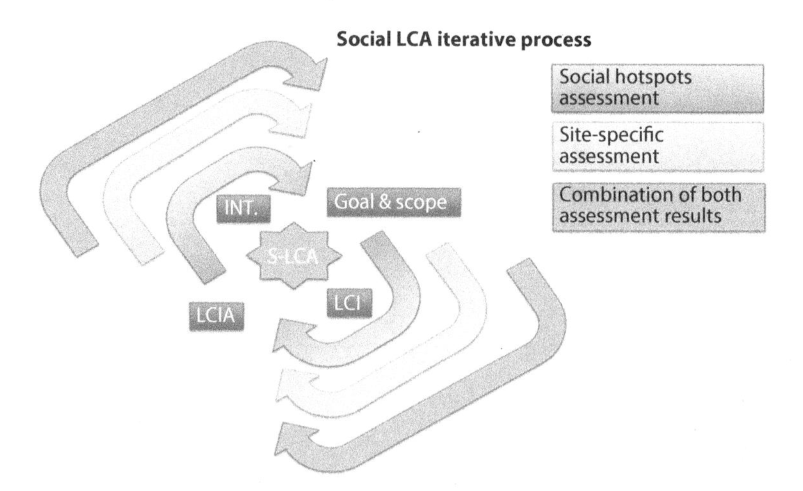

Figure 7.3 Social LCA is an iterative process.

A social hotspots assessment itself (also called a generic or scoping assessment) will go through its own loop of Goal and Scope, Inventory analysis, Impact assessment, and interpretation. In a system such as the social hotspots database, this loop is automated. Once results of the social scoping assessment are obtained, the goal and scope can be revised and, if part of the scope, further data collection activities may take place. After conducting the life cycle inventory analysis of the site-specific investigation, goal and scope may be revised again based, for instance, on data availability.

7.4.2.1 Goal and Scope

Several goals may be sought from carrying a SLCA. It may be that a new product or building is being designed, that a consumer good is being assessed for improvement of its social responsibility, or that an assessment is being carried out to inform public policy. Studies may be performed to investigate the potential negative impacts or they may be conducted to provide a holistic picture of the positive and negative impacts generated by the production activities. The intended use, the type of stakeholders to be reached by the results, and the detail of what issues need to be assessed will influence the planning of the assessment (UNEP/SETAC 2009). The availability of information about the supply chain of the product, or lack thereof, will affect the design of the study. In either case, a functional unit will be determined. The choice of the functional unit may also be influenced by the type of study (generic, site-specific) and the availability of models (IO, Unit process).

The product supply chain is unknown

When an SLCA is carried for a design or product development project, very little information may be available on the product supply chain, thus a site-specific assessment (and the integration of site-specific indicators) is irrelevant.

The most appropriate level of assessment may be a social hotspots assessment that can be carried on one or several alternatives. Based on the results, more detailed information may be collected (for instance can some of the inputs be sourced from fair trade certified sources). A common mistake consists in integrating site-specific indicators to SLCA studies intended for product development (for example Sandin *et al* 2011).

Information about the product supply chain (or product category) is known

In the case that at least some product supply chain information is known, site-specific collection activities may be planned to investigate parts of the product system. Often enough, the composition of a product supply chain changes. Whether it is because a certain input is only available at certain time of the year in a given country (i.e. in the case of vegetables and fruits) or because price changes affect product formulation (i.e. in the case of home and personal care products), supply chains may not be as stable as foreseen. When a site-specific investigation is planned, site-specific indicators may be integrated to the study design. In order to select indicators, it is useful to gather information on social responsibility systems that may be used in the assessed supply chain. Many retailers and brands now use data platforms to share audit information (i.e. Sedex, Fair Factory Clearinghouse, Ecovadis). It may be beneficial to use the same indicators in the SLCA that are used during the audit process in order to mine existing resources.

7.4.2.2 Scoping Decisions

The determination of the depth and breadth of the study will be influenced by the assessed goal(s). Scoping decision will have to be taken, for instance, on the product system, the stakeholder groups to be investigated, the geography, the subcategories of assessment, and the indicators.

At the level of the product system, decisions need to be made on the inclusion or exclusion of:

Retail
Transport
Infrastructure components
Services (such as financial or legal services)
Use phase
End of life
(...)

In the case of buildings, a more detailed use phase investigation may be planned as part of the scope. The end of life may be excluded in the case of a study aiming at assessing improvement opportunities in the fresh flower supply chain. Decisions about inclusions or exclusions may also be reviewed after a social hotspots assessment. Services might have been excluded from the scope but re-included because of contributing to a large share of the worker hours.

As part of the scoping exercise, decisions will be taken regarding the inclusion/exclusion of stakeholder groups. For instance, it can be decided to focus on a specific stakeholder group (the worker for instance). Dreyer provides a good example of such a study design (Dreyer 2010). Some stakeholder groups may also not be relevant to investigate for some part of the supply chain (for instance, local community in the case of services or transport activities).

Specifying the geography of the functional unit is also something that is strongly recommended. Orange juice intended for the U.S. market does not use oranges from the same sourcing countries as orange juice for sale in the European market. Trade data can be collected to inform subsequent geographical decisions or a global input output model can be used.

Regarding choices of impact subcategories, it was mentioned before that, some may not bring value to a scoping assessment. Different sets of subcategories may be used for different parts of the supply chain. For example, local community, indigenous rights may not be relevant to study for the manufacturing of plastic toys in a big city such as Shanghai, China, thus should be excluded for this phase; but re-included when assessing the paper used in packaging.

Decision on subcategories inclusion will be determinant for the choice of indicators. The type of study (generic or site-specific) will also influence indicator's choice. Finally, indicator schemes in use in the studied supply chain (Global Social Compliance Programme questionnaire for instance), may inform the selection.

The SLCA Guidelines highlight the necessity to document and be transparent about all the scoping decisions for a study to be valid.

7.4.2.3 Life Cycle Inventory for SLCA

Life Cycle Inventory consists in collecting and organizing the data, modeling the systems, and obtaining LCI results. The goal and scope of the study provide an initial plan for conducting the inventory phase of the SLCA. The SLCA Guidelines describe multiple steps associated with the Life Cycle Inventory. Without databases, information needs to be collected on activity variables such as worker hours and social hotspots. As additional steps, more detailed data and site-specific information may be collected if required. The inventory is undoubtedly the most time consuming task of the SLCA. The more it can be automated, the more time and cost efficient studies can be. Data collection activities are being carried out using different methods depending whether the study is generic or site-specific. Social scoping assessments may be carried using secondary databases of information, peer reviewed literature, internet search and expert interviews. Site-specific assessment will require questionnaires and audits.

7.4.2.4 Social Life Cycle Impact Assessment

The SLCA Guidelines present a LCIA framework structured as a set of impact categories being aggregates of subcategories, which in turn are aggregates of inventory indicators. It specifies that the characterization models used for aggregating inventory information needs to be formalized. Aggregation may take place through summarizing qualitative information or by adding up quantitative information. The social inventory information may, in other cases, require a scoring system to facilitate the meaning assessment, relating the information to Performance Reference Points. Franze and Ciroth (2011) used the latter approach in their laptop case study. Performance Reference Points are additional information used in characterization models. Performance reference points may be internationally set thresholds, goals or objectives according to conventions and best practices, etc. (UNEP/SETAC 2009).

Scoring systems provide a way to handle the distribution of positive and negative impacts in relation to stakeholder needs and context, which is an important task of the impact assessment. The effect of potentially improved or worsened social conditions may have a significant effect on the result.

Another way of aggregating the inventory information is through life cycle attribute assessment (LCAA), a technique introduced by Greg Norris in 2006 (Norris 2006), LCAA calculates the share of relevant activity across a life cycle, which has attributes of interest. This generates results as such: 80% of the worker hours of the life cycle of the product are known to be child labor free (Benoît *et al* 2010).

Finally, LCIA may involve causal chain modeling through impact pathways. The development of impact pathways is a topic for further research since very few pathways are currently well defined in the social science literature. The use of impact pathways in SLCA also raises issues of concern, some of which are discussed in Jorgensen *et al* 2010. In particular, stakeholders in the case of no-go issues such as forced labor, or the worst forms of child labor, may question the relevance to further model the consequences on autonomy or other appropriate Area of Protection.

A key characteristic of SLCA is its aim to capture positive impacts. In that regard, SLCA occupies a special niche among sustainability assessment techniques.

7.4.2.5 Interpretation of SLCA

Interpretation is a key step in a LCA study. For example, when social hotspots assessments are performed, a very high number of unit processes or CSS may be identified to be at high risk. Strategies need to be implemented to identify the most significant issues.

Associated with the interpretation phase is also the evaluation of the completeness and consistency of the SLCA. It may be that, following Interpretation, an additional assessment loop needs to be conducted in order to provide greater completeness or to make boundaries more consistent.

Interpretation also consists in presenting results. Results can be presented by format decided by the character of the results, in respect to analytical purposes, or by their robustness. Results can be presented interactively using web based software or via a report or slides.

7.4.3 SLCA and other Key Social Responsibility References and Instruments

Several key social responsibility references and instruments have already been mentioned, including the international policy frameworks (i.e. the UN Protect, Respect and Remedy framework, UN and ILO International conventions), SR Implementation Guidelines (i.e. ISO 26000 Guidance document on Social Responsibility), as well as Auditing and Monitoring Frameworks (i.e. the Global Social Compliance Programme, Better work Programme).

The SLCA technique makes use of the modeling capabilities and systematic assessment processes of LCA for the analysis of the positive and negative social impacts engendered by production activities. The impacts are largely defined by the international community through its policy frameworks and other social responsibility references and in respect to best available sciences (top down approach). Jørgensen captures the importance of deontological ethic for SLCA when he states that some of the subcategories are expected to be assessed by principle rather than by consequence (Jørgensen *et al* 2008). This calls for the necessity to situate SLCAs in the greater societal context of the Social Responsibility "movement" and in particular the international policy framework.

Pragmatically, conducting SLCAs in a timely and cost-efficient manner necessitate knowing about and making use of all resources available.

Seven main types of references and instruments have been identified as relevant to social sustainability assessment (Benoît and Vickery Niederman 2010): International Policy Frameworks (i.e. International Conventions), Codes of Conduct and Principles (i.e. company own codes of conduct, Caux Round Table principles for business), Sustainability Reporting Frameworks (i.e. GRI), SR Implementation Guidelines (i.e. ISO 26000), Auditing and Monitoring Frameworks (i.e. Global Social Compliance Programme), and Financial Indices (i.e. Dow Jones Sustainability Indexes). In addition, three social assessment methods are relevant to consider when planning social sustainability assessment. These include Social Impact Assessment, Human Rights Impact Assessment and Value Chain Analysis.

A detailed description of the types of references and instruments can be found in Benoît and Vickery-Niederman 2010. The references and instruments can be classified by their relevance for different phases of SLCA.

The references are relevant to the Goal and Scope phase if they inform decisions relative to the assessment framework and the identification of indicators. The International policy frameworks constitute the foundation for all social responsibility initiatives, references instruments and techniques including SLCA (UNEP/SETAC 2009). To be

Table 7.3 Social Responsibility (SR) Instruments, References and Methods Relevant for Each Phase of SLCA.

SLCA phase	Types of instrument, reference or method
Goal and Scope/ Determination and definition of subcategories and indicators	International Policy Frameworks, Codes of Conduct and Principles, Sustainability Reporting Frameworks, SR Implementation Guidelines
Life Cycle Inventory	Sustainability Reporting Frameworks, Auditing and Monitoring Frameworks and Financial Indices, Social Impact Assessment, Human Right Impact Assessment
Life Cycle Impact Assessment	International Policy Frameworks
Interpretation	International Policy Frameworks, SR Implementation Guidelines, Sustainability Reporting Frameworks

relevant to the Life Cycle Inventory phase, the instruments and references need to offer data collection methods or be a source of data. Instruments and references are meaningful to LCIA when they can provide Performance Reference Points. Finally, references and instruments are useful at the interpretation phase if they can inform the identification of significant issues or can be a useful tool to the presentation of results and the drafting of recommendations.

The development of databases for SLCA was recommended by the Guidelines (UNEP-SETAC, 2009) and many authors of SLCA articles (for example: Dreyer 2010, Hauschild 2008, Jørgensen 2010). The availability of databases for environmental LCA has dramatically increased the usability of the technique and the same is expected for SLCA.

Regarding data, SLCA faces very much the same challenges as its environmental counterpart. Data may be sparse and site-specific data collection activities are costly and time consuming. In environmental LCA averages are often used to compile environmental inventories. When measuring environmental impacts that are local *by definition*, such as soil erosion, this reduces the validity of the assessment (Benoît and Vickery Niederman, 2010).

In SLCA, the geographic resolution of data presents an even greater challenge. This represents one of the reasons why Social Hotspots Assessment is useful to SLCA. Also called generic analysis or social scoping assessment, Social Hotspots Assessment is a screening device that allows users to narrow in on the locations, sectors, inputs or unit processes with an increased risk for social violations or opportunities for positive social actions.

The use of country level information is sometimes the most appropriate in SLCA studies. Ekvall argues that "such information has proven to be relevant at all levels of decision making. It has, for example, many times affected decisions of consumers as individuals or groups" (Ekvall 2010).

The Social Hotspots Database (SHDB) is being compiled as a tool for generic analysis (Benoît *et al* 2012). It is the first comprehensive data source for SLCA. It consists of a three-layer system. The first layer is composed of a global input output model derived by New Earth from the Global Trade Analysis Project (GTAP). The first layer provides global IO modeling capabilities providing estimates on the share of sector specific economic activities happening in different countries and regions of the world in relation to a quantity of country specific sector economic output generated (the "functional unit" i.e. 1 million dollars of US Dairy products). GTAP provides information according to a matrix consisting of 57 sectors and 113 countries or regions.

The second layer provides an estimate of worker hours. Following recommendations from the Guidelines (UNEP/SETAC 2009), it is proposed by New Earth as a key activity variable and one of three criteria to determine where in a supply chain the greatest risks or opportunities may lie (i.e. hotspots), due to the intensity of the labor hours associated with a particular unit process or CSS. It utilizes Life Cycle Attribute Assessment, as a mean to aggregate social (and environmental) attributes throughout the supply chains using activity variables. The two additional criteria consist of an assessment of the severity of the risk/opportunity (from very high to low) and the gravity of the issue (also on a scale).

The third layer consists of social data that are grouped into 20 social themes (that correspond to subcategories of assessment) covering 191 countries and 57 economic sectors. One or more indicator is used to investigate every theme. Currently, the database integrates over 50 indicators. Whenever possible, triangulation of data is implemented and different sources of information are compared across social themes to identify differences and discrepancies in data and data interpretation, thus decreasing uncertainty in the results. Almost 200 different sources of data have been used to build the existing tables.

The SHDB is data driven; it incorporates the best *available* information. SHDB data on social themes may not be available for every country, nor for every sector, simply because they do not exist in the publicly available literature. In that sense, the database is a meta-analysis of the best international data available.

The project uses criteria to guide data collection (the order does not represent a hierarchy). Data is incorporated into the database on the basis of:

1. Comprehensiveness (data is available for a large number of countries and/or sectors),
2. Meaningfulness of the indicators (ability of the indicators to capture the theme investigated),
3. Legitimacy of the data source (collected and distributed by well-recognized organizations),
4. Quality of the data (minimizing uncertainty),
5. Quantitative information (often preferred when available)

SHDB inventory data is then processed by characterization models. In the Guidelines SLCIA characterization models are defined as the "formalized…operationalization of the social and socio-economic impact mechanisms. They may be a basic aggregation step, bringing text or qualitative inventory information together into a single summary, or summing up quantitative social and socio-economic inventory data within a category". The Guidelines specifies that "characterization models may also be more complex, involving the use of additional information" (UNEP/SETAC 2009).

In the SHDB, the characterization models bring the inventory information together by assigning a degree of risk or opportunity to the data (Low, Medium, High or Very High). A distribution of the global data is utilized for the majority of the tables as a basis for the characterization model. For some tables, a review of the literature provided more accurate interpretations of the risk. An understanding of global or regional data averages, or access to expert knowledge on the subject, is very desirable in ranking the indicators as low to very high probability for a social issue to occur.

The SHDB allows users to screen for social risk/opportunities,,and then focus data gathering efforts on their facilities in the country-specific sectors identified as most at risk. In addition, users can initiate social development projects in these same countries and thus focus improving the social conditions of production in areas with the most need. The information is meant to be a basis for more detailed data collection activities. Instead of using a general index such as the Human Development Index, the SHDB allows a greater level of refinement in the assessment of the occurrence of the potential social impacts. It would be very counterproductive to use a device such as the Social Hotspots Database to make sourcing decision. Instead, it should be used to ask the right questions to suppliers, gather more data, and make decisions about conducting audits.

Table 7.4 Uses of the Social Hotspots Database.

Uses of SHDB information	Prioritize data collection activities
	Screen for social risks and opportunities
	Put site-specific assessment results in perspective
	Prioritize supply chains' improvement initiatives or philanthropy
	Understand better the social conditions in a country/sector
Uses of SHDB LCAA model	Express the % of a supply chain which possesses a risk/opportunity or attribute of interest
	Assess supply chain labor intensity by country specific sector
	Model supply chains by country specific sector

7.5 Research Needs for LCSA Methodology

The LCSA distinguishes itself from other approaches for being an integrative analysis, covering a wide range of methods and tools, addressing the full range of scales from local to global, and linking knowledge of different disciplines and normative positions. For small scale decisions on products and technologies, LCA has been developed and it remains a valuable method in the LCSA framework. Moreover, the developments that have been occurring in the last years outline a methodology that goes beyond the ISO-LCA and moves towards those dimensions we pointed out as relevant for a LCSA: broadening and deepening. On the deepening side, sophistication in modeling for example is achieved in many ways: consequential LCA, introduction of time dimension, hybrid approaches combining input-output and LCA, and the numerous developments occurred in impact assessment, ranging from regionalization to the development of new characterization methods.

Approaches aimed at making LCA broader, going beyond the focus on environmental aspects, are represented by the development of LCC and SLCA, among others. Even if they show a different degree of maturity, and in particular SLCA is still in its infancy, notable developments are expected in the next years. In fact, since these two methodologies are explicitly required, together with LCA, in the calls of the European Research 7th Framework Programme for evaluating the sustainability assessment of technologies, we could expect that the case studies developed during the projects will lead to further develop the methods.

These developments have characterized the story of LCA of the last ten years (2000-2010), and represent an attempt to better model the system, reduce uncertainty, collect more representative data, define scenarios and include more mechanisms in the analysis than the environmental and technological ones. They are a good example of interdisciplinarity, a peculiarity of the LCA framework which historically works with the support of other models and disciplines, as clearly demonstrated in the impact assessment phase. However, the integration and/or combination with other methods and models, especially for the modeling phase, are questioned by many researchers.

An explanation could be that LCA, as conceived now, has been developed for certain types of applications and that the expansion of the methodology towards a broader and deeper method could violate its inherent principles. Indeed, this concern could be justified by the fact that LCA was very criticized in the past, and only with standardization it gained again a reputation. Therefore, any major change of LCA could endanger again its credibility. On the other hand, others could assert that the LCA framework has been conceived flexible enough to allow the combination and/or integration with other models, and thus it is part of its character to grow and to further expand.

The contribution of other disciplines is certainly fundamental in advancing the analysis and a necessary ingredient for the LCSA framework, since we cannot find all the answer in LCA as presently available. The scientific community has to fill the identified gaps in theory, practice and use within the context of sustainability support, carrying out its principal task of continuously investigating new lines of developments, testing their scientific soundness, selecting the most promising approaches and preparing the next steps of development. Research for making the LCSA framework well-developed and fully applicable is highly demanding and requires intervening at least at three levels (Guinée *et al* 2009):

- Broadening and deepening the scope of indicators (sustainability indicators for LCSA);
- Deepening the scope of mechanisms;
- Cross-cutting research for integration.

For each of these main headings, a number of research lines can be defined. We would like to point out those which we consider more urgent to address and whose results could benefit for the others.

Aligning environmental with economic and social indicators

Indicators for sustainability have to satisfy many and different requirements, among which an important one is their mutual alignment: they should refer to the same decision situation and be based on the same empirical relations and assumptions. This process of alignment involves not only the extension to cover more issues than the environmental ones and other level of analysis, but also the modeling side. In fact, aspects like their place in the causal chains and their mutual independence, their time specification and spatial differentiation are aspects that need to be dealt with to develop robust indicators and metrics for LCSA.

Framing the question

How to properly frame the questions and consequently how to better link questions and models is an important field of research. In fact, the depth and the breadth of any evaluation can differ considerably depending on the goal and scope of the study and errors made in this phase have strong consequences on the results (Fullana *et al* 2011).

This research line requires the development of practical guidelines which support practitioners in defining the following relevant aspects: identify what the problem we are trying to tackle is exactly, what the derived questions are, what the technological options are, what the scale of the expected changes is, what the time frame of the question is, if a ceteris paribus assumption may hold or not, if the system analyzed is replacing another system at a small scale or if it is expected that the technology used in the new system will probably expand to many more applications on a larger scale, etc. It is the sum of all these answers that determines which methodological choices are relevant, and thus, which method is more appropriate.

Modeling options for meso-level and economy-wide applications

This research line is aimed at investigating how to quantitatively model mechanisms for each level of LCSA, i.e. product, sector and economy-wide. In this regard, the analysis may develop in two directions:

- Incorporating more mechanisms in the analysis, either endogenised or through a set of more softly linked heterogeneous models;
- Specifying scenarios, with causalities modeled partially within the scenario framework. For questions on larger scale technology options, as on new energy supply systems, such scenario based modeling seems most adequate.

Practical models and tools need to be identified specifically for each level of LCSA, and one of the main results of this research line should be the provision of practical tools and guidelines for practitioners on how to operate these models.

Together with these main research lines, there are also cross-issues which need to be investigated, among which the treatment of the uncertainty and the development of simplified approaches deserve particular attention. As far as the uncertainty is concerned, we should consider that it is inherent to any evaluation and as such cannot be eliminated but only managed and hopefully reduced. Moreover, for complex systems the verification and validation of empirical models is impossible due to the high numbers of parameters involved and interconnections. The challenge for the LCSA framework is thus to broaden and deepen the analysis while managing or counteracting the resulting increase in complexity. Moreover, since complexities are at many levels (the real world in which we live, the understanding of the main relations in the models and their modeling), simplifications will be necessary, without leaving relevant mechanisms and aspects out of the analysis.

References

Andersson K, Eide MH, Lundqvist U and Mattsson B (1998) The Feasibility of Including Sustainability in LCA for Product Development *J Clean Prod* 6(3-4):289-298.

Assefa G and Frostell B (2007) Social Sustainability and Social Acceptance in Technology Assessment: A Case Study of Energy Technologies *Technology in Society* 29:63-78.

Benoît C and Mazijn B (2009) (eds) Guidelines for Social Life Cycle Assessment of products. UNEP/SETAC Life Cycle Initiative, ISBN: 978-92-807-3021-0, France.

Benoît C, Norris G, Valdivia S, Ciroth A, Moberg Å, Bos U, Prakash S, Ugaya C and Beck T (2010) *Int J Life Cycle Assess* 15(2):156-163.

Benoît C and Vickery-Niederman G (2010) Social Sustainability Assessment Literature Review, White Paper #102, The Sustainability Consortium, http://www.sustainabilityconsortium.org/wp-content/themes/sustainability/assets/pdf/whitepapers/Social_Sustainability_Assessment.pdf.

Benoît C, Vickery-Niederman G, Valdivia S, Franze H, Traverso M, Ciroth A and Mazijn B (2011) Introducing the UNEP/SETAC Methodological Sheets for Subcategories of Social LCA *Int J Life Cycle Assess* 16:682–690.

Benoît-Norris C, Aulisio D, Norris GA, Hallisey-Kepka C, Overraker S and Vickery Niederman G(2011) A Social Hotspots Database for Acquiring Greater Visibility in Product Supply Chains: Overview and Applications to Orange Juice, in proceedings *Towards Life Cycle Sustainability Management*, Springer, M. Finkbeiner, ed, pp. 53-64.

Benoît-Norris C, Aulisio Cavan D and Norris G (2012) Identifying Social Impacts in Product Supply Chains: Overview and Application of the Social Hotspot Database *Sustainability 4*(9): 1946-1965; doi:10.3390/su4091946.

Bowen H (1953) Social Responsibilities of the Businessman, New York, Harper & Row.

Bruins RJF, Munns WR, Botti SJ, Brink S, Cleland D, Kapustka L, Lee D, Luzadis V, McCarthy L.F, Rana N, Rideout D.B, Rollins M, Woodbury P and Zupko M (2009) A New Process for Organizing Assessments of Social, Economic, and Environmental Outcomes: Case Study of Wildland Fire Management in the USA **Integrated** *Environ* **Assess** *Manag* 6(3):469-483.

Capron M, Quairel-Lanoizelée F, and Turcotte M-F (2010) ISO 26 000: une norme hors norme? Paris, Economica.

Ciroth A and Franze J (2011) LCA of an Ecolabeled notebook. Consideration of Social and Environmental Impacts along the Entire Life Cycle, ISBN 978-1-4466-0087-0, Berlin.

COM (2008) Communication from the Commission to the European Parliament, the Council, the European Economic and Social Committee and the Committee of the Regions on the Sustainable Consumption and Production and Sustainable Industrial Policy Action Plan, COM (2008) 397/3.

COM (2005) Communication from the Commission to the European Parliament, the Council, the European Economic and Social Committee and the Committee of the Regions. Thematic Strategy on the Sustainable Use of Natural Resources. COM(2005)670.

COM (2004) Communication from the Commission to the European Parliament, the Council, the European Economic and Social Committee and the Committee of the Regions Stimulating Technologies for Sustainable Development: An Environmental Technologies Action Plan for the European Union. COM(2004) 38 final.

COM (2003) Communication from the Commission to the European Parliament, the Council, the European Economic and Social Committee and the Committee of the Regions. Integrated Product Policy. Building on Environmental Life-Cycle Thinking. COM(2003)302 final.

Curran, M.A.; Mann, M. and Norris, G. (2005) International Workshop on Electricity Data for Life Cycle Inventories *J Cleaner Prod*, Vol. 13, No. 8, p. 853.

Dreyer L, Hauschild MZ and Schierbeck J (2006) A Framework for Social Life Cycle Impact Assessment *Int J Life Cycle Assess*, Vol.11, No. 2, P. 88-97.

Dreyer LC, Hauschild MZ and Schierbeck J (2010) Characterisation of Social Impacts in LCA *Int J Life Cycle Assess* 15:247-259.

EC (2009) Directive 2009/125/EC of the European Parliament and of the Council of 21 October 2009 Establishing a Framework for the Setting of Ecodesign Requirements for Energy-Related Products. Official Journal of the European Union, EC, L285. 31.10.2009: 10-35.

EC (2006) Directive 2006/12/EC of the European Parliament and of the Council of 5 April 2006 on waste. Official Journal of the European Union, EC, L114. 27.4.2006: 9-21.

Elkington J (1998) Partnerships from "*Cannibals with Forks: The Triple Bottom Line of 21st-Century Business" Environ Quality Management* 8(1):37-51.

Finkbeiner M, Schau EM, Lehmann A and Traverso M (2010) Towards Life Cycle Sustainability Assessment *Sustainability* 2(10): 3309-3322.

Franze J and Ciroth A (2011) A Comparison of Cut Roses from Ecuador and the Netherlands *Int J Life Cycle Assess* 16(4):366-379.

Friedrich R (2011) *J Ind Ecol* 15(5):668.

Fullana P, Puig R, Bala A, Baquero G, Riba J and Raugei M (2011) From Life Cycle Assessment to Life Cycle Management: A Case Study on Industrial Waste Management Policy Making *J Ind Ecol* 15(3): 458-475.

Girod B Haan P and Scholz RW (2010) *Int J Life Cycle Assess* 16(1):3.

Graedel TE and Klee RJ (2002) *Environ Sci Technol*, Vol. 36, No. 4, p. 523.

Graedel TE and van der Voet E (eds) (2010) *Linkages of Sustainability*, the MIT Press Cambridge, Massachusetts London, England.

Guinée JB, Gorrée M, Heijungs R, Huppes G, Kleijn R, de Koning A, van Oers L, Wegener Sleeswijk A, Suh S, Udo de Haes HA, de Bruijn JA, van Duin R and Huijbregts MAJ (2002) (eds) *Handbook on life cycle assessment: Operational guide to the ISO standards. Series: Eco-efficiency in industry and science*, Dordrecht, Kluwer Academic.

Guinée JB and Heijungs R (2011) *J Ind Ecol*, Vol. 15, No. 5, p.656.

Guinée JB, Heijungs R, Huppes G, Zamagni A, Masoni P, Buonamici R, Ekvall T and Rydberg T (2011) *Environ Sci Technol*, Vol 45, p. 90.

Guinée JB, Huppes G, Heijungs R and van der Voet E (2009) Research Strategy, Programmes and Exemplary Projects on Life Cycle Sustainability Analysis (LCSA), Technical Report of CALCAS Project, http://www.calcasproject.net.

Hauschild MZ, Dreyer LC and Jørgensen A (2008) CIRP Annals - Manufacturing Technology, Vol. 57, pp. 21-24.

Heijungs R, Guinée JB and Huppes G (2009) A Scientific Framework for LCA, Technical Report of CALCAS project, http://www.calcasproject.net.

Heijungs R, Huppes G, and Guinée JB (2010) *Polymer Degradation and Stability* 95 (3):422.

Heijungs R, Huppes G, Guinée JB, Masoni P, Buonamici R, Zamagni A, Ekvall T, Rydberg T, Stripple H, Rubik F, Jacob K, Vagt H, Schepelmann P, Ritthoff M, Moll S, Suomalainen K, Ferrão P, Sonnemann G, Valdivia S, Pennington DW, Bersani R, Azapagic A, Rabl A, Spadaro J.; Bonnin D.; Kallio A.; Turk D and Whittall J (2007) Scope of and Scientific Framework for the CALCAS Concerted Action, Technical Report of CALCAS Project, http://www.calcasproject.net.

Hunkeler D, Lichtenvort K and Rebitzer G (2008) (eds) *Environmental Life Cycle Costing*, SETAC-CRC, Pensacola.

Hunkeler D and Rebitzer G (2005) The Future of Life Cycle Assessment *Int J Life Cycle Assess* 10(5): 305.

Huppes G and Ishikawa, M (2007) Sustainable Futures: The Rationale for the Working Group on Modeling and Evaluation for Sustainability *J Ind Ecol* 11(3):7-10.

Huppes G, van Rooijen M and Kleijn R (2004) Life Cycle Costing and the Environment, With Dutch Summary, Report VROM-DGM commissioned by the Ministry of the Environment for RIVM Expertise Centre LCA, Zaaknummer 200307074, http://www.rivm.nl/milieuportaal/images/Report%20LCC%20April%20%202004%20final.pdf.

ISO (2006a) ISO 14040:2006: Environmental Management—Life Cycle Assessment—Principles and Framework, Geneva, Switzerland.

ISO (2006b) ISO 14044:2006: Environmental Management—Life Cycle Assessment—Requirements and Guidelines, ISO, Geneva, Switzerland.

ISO (2010) International Organization for Standardization, ISO 26000 Guidance on Social Responsibility. Geneva, International Organization for Standardization.

Jørgensen A, Le Bocq A, Nazarkina L and Hauschild M (2008) Methodologies for Social Life Cycle Assessment *Int J Life Cycle Assess* 13(2):96-103.

Jørgensen A, Lai L and Hauschild M (2010) Assessing the Validity of Impact Pathways for Child Labour and Well-Being in Social Life Cycle Assessment_*Int J Life Cycle Assess* 15(1):5-16, 2010.

JRC-IES (2010a) ILCD Handbook: General Guide for Life Cycle Assessment - Provisions and Action Steps, European Commission, Ispra, Italy, http://lct.jrc.ec.europa.eu/pdfdirectory/ILCD-Handbook-General-guide-for-LCA-PROVISIONS-online-12March2010.pdf, 2010.

JRC-IES (2010b) ILCD Handbook: Analysis of Existing Environmental Impact Assessment Methodologies for Use in Life Cycle Impact Assessment, European Commission, Ispra, Italy.

Klöpffer W (2008) Life Cycle Sustainability Assessment of Products *Int J Life Cycle Assess* 13(2): 89-95.

Norris GA (2006) Social Impacts in Product Life Cycles - Towards Life Cycle Attribute Assessment *Int J Life Cycle Assess* 11, Issue Supplement 1:97-104.

Ruggie J (2008) Protect, Respect and Remedy: A Framework for Business and Human Rights, Report to the Human Right Council, http://www.businesshumanrights.org/SpecialRepPortal/Home/ReportstoUNHumanRightsCouncil/2008.

Saling P, Kicherer A, Dittrich-Kriimer B, Wittlinger R, Zombik W, Schmidt I, Schrott W and Schmidt S (2002) Eco-efficiency analysis by BASF: The Method *Int J Life Cycle Assess* 7(4):203-218.

Saling P, Maisch R, Silvani M and König N (2005) Assessing the Environmental-Hazard Potential for Life Cycle Assessment, EcoEfficiency and SEEbalance *Int J Life Cycle Assess* 10(5): 364-371.

Sandin G, Peters G, Pilgård A, Svanström M, and Westin M (20110 Integrating Sustainability Considerations into Product Development: A Practical Tool for Prioritising Social Sustainability Indicators and Experience from Real Case Applications, in LCM proceedings Towards Life Cycle Sustainability Management, M. Finkbeiner, ed, Springer, pp. 3-14

Schepelmann P, Ritthoff M, Santman P, Jeswani H and Azapagic A (2008) D10 Report on the SWOT analysis of concepts, methods and models potentially supporting LCA, Technical Report of CALCAS project, http://www.calcasproject.net.

Searchinger T, Heimlich R, Houghton RA, Dong F, Elobeid A, Fabiosa J, Tokgoz S and Hayes D (2008) Use of U.S. Croplands for Biofuels Increases Greenhouse Gases Through Emissions from Land-Use Change *Science* 319(5867):1238-1240.

Tukker A (2011) Harmonizing Science and Policy Programs for a Decent and Sustainable Life for All by the Mid-Millennium *J Ind Ecol* 15(5):650-652.

UNEP/SETAC (2009) Guidelines for Social Life Cycle Assessment of Products, C. Benoît, and B. Mazijn, eds. United Nations Environment Programme, Paris.

Upham P (2000) An Assessment of The Natural Step Theory of Sustainability *J Clean Prod* 8(6):445-454.

Vanclay F (2009) "Conceptual and methodological advances in social impact assessment", in H.A. Becker and F. Vanclay (eds) *The international handbook of social impact assessment – conceptual and methodological advances*, Boston, Edward Elgar.

Verones F, Hanafiah MM, Pfister S, Huijbregts MAJ, Pelletier GJ and Koehler A (2010) *Environ Sci Technol* 44(24):9364.
Weidema BP (2011) *J Ind Ecol* 15(5):658.
Weidema BP (2006) *Int J Life Cycle Assess* 11, Special Issue No. 1, p. 89.
Zamagni A, Amerighi O and Buttol P (2011) *Int J Life Cycle Assess*, 16(7):596.
Zamagni A, Buonamici R, Buttol P, Porta PL and Masoni P (2009) Main R&D Lines to Improve Reliability, Significance and Usability of Standardised LCA, Technical Report of the CALCAS project. http://www.calcasproject.net.
Zhang Y, Baral A and Bakshi BR (2010) *Environ Sci Technol* 44:2624.

Chapter 7 Exercises

1. Short-Term and Long-Term Sustainability
   This chapter refers to weak versus strong sustainability. This concept has also been referred to elsewhere as short-term and long-term sustainability. What is your understanding of these terms? Provide an example of an activity that could be considered as weak sustainability and one that could be considered strong. Why is it important to identify these different strategies?
2. Framework for LCSA
   This chapter refers to a suggested framework for sustainability, expressed by the symbolic equation

   $$\text{LCSA} = \text{LCA} + \text{LCC} + \text{SLCA},$$

   where LCC and SLCA are Life Cycle Costing and Social Life Cycle Assessment, respectively.
   a) What is your understanding of the relationship between these three techniques?
   b) Why is the LCSA equation described as "symbolic?"
   c) What is are the similarities and differences between the three variables in the LCSA equation and the standard dimensions of sustainability (environment, economy, and society)?
3. Social LCA
   a) How is conducting an SLCA similar to conducting an LCA? How do they differ?
   b) Similar to environmental indicators in LCA, the selection of social indicators must be relevant to the study goal and scope. How could the relevancy of social indicators be determined?
   c) What type of inventory data might be needed to conduct an SLCA?

8

Resources for Conducting Life Cycle Assessment

Books

LCA Compendium: The Complete World of Life Cycle Assessment (2014) Walter Klöpffer and Mary Ann Curran (eds); Springer; ISSN: 2214-3505.

Life Cycle Assessment (LCA) (2014) Walter Klöpffer and Birgit Grahl, Wiley-VCH; ISBN 978-3527329861; 440pp.

Environmental Life Cycle Assessment: Measuring the Environmental Performance of Products (2014) Rita Schenck and Philip White (eds) American Center for Life Cycle Assessment; ISBN-13: 978-0988214552; 322pp.

The Life Cycle Assessment Handbook: A Guide for Environmentally Sustainable Products (2012) Mary Ann Curran (ed); Scrivener-Wiley; ISBN 978-1118099728; 625pp.

Lifecycle Assessment: An Introduction for Students (2011) prepared by Jeremy Faludi

http://sustainabilityworkshop.autodesk.com/sites/default/files/core-section-files/lcaprimer_autodesk-sworkshop_final.pdf

Life Cycle Assessment: Principles, Practice and Prospects (2009) Ralph Horne, Tim Grant and Karli Verghese (eds) CSIRO Publishing; 192pp.

Life Cycle Assessment: Principles and Practice (2006) EPA/600/R-06/060, US Environmental Protection Agency, Office of Research and Development, Washington, DC;

available on-line, www.epa.gov/ORD/NRMRL/std/lca/lca.html.

Environmental Life Cycle Assessment of Goods and Services: An Input-Output Approach (2006) Hendrickson, Chris T., Lester B. Lave, H. Scott

Matthews, Arpad Horvath, Satish Joshi, Francis C. McMichael, Heather MacLean, Gyorgyi Cicas, Deanna Matthews and Joule Bergerson, Resources for the Future Press; ISBN 978-1933115238; 272pp.

Why Take a Life Cycle Approach? (2004) Prepared for the UNEP/SETAC Life Cycle Initiative, United Nations Publications ISBN 92-807-24500-9; 24 pp.

The Hitch Hiker's Guide to LCA (2004) Henrikke Bauman and Anne-MarieTillman; Studentlitteratur AB; ISBN 978-9144023649; 544pp.

Handbook on Life Cycle Assessment: Operational Guide to the ISO Standards (2002) Jeroen Guinée (ed.), Springer; ISBN 978-1402002281; 692pp.

The Computational Structure of Life Cycle Assessment (2002) Reinout Heijungs and Sangwon Suh; Kluwer Academic Publishers; ISBN 978-1-4020-0672-2; 243pp.

Life Cycle Assessment: Inventory Guidelines and Principles (1992) EPA/600/R-92/245, US Environmental Protection Agency, Office of Research and Development, Washington, DC.

Organizations

European Platform on Life Cycle Assessment

http://eplca.jrc.ec.europa.eu/

The European Platform on LCA is a project of the European Commission, carried out by the Commission's Joint Research Centre (JRC), Institute for Environment and Sustainability (IES) in collaboration with DG Environment Directorate Green Economy. It has facilitated the development of the European reference Life Cycle Database (ELCD), the International reference Life Cycle Data System (ILCD) Handbook, and the Life Cycle Data Network (LCDN).

Publications:

- International Reference Life Cycle Data System (ILCD) Handbook (2012) EUR 24892 EN; 72pp.
- General Guide for Life Cycle Assessment - Detailed Guide (2010) EUR 24708 EN; 417pp.
- General Guide for Life Cycle Assessment - Provisions and Action Steps (2010) EUR 24378 EN; 163pp.
- Specific Guide for Life Cycle Inventory Data Sets (2010) EUR 24709 EN; 142pp.
- Analysis of Existing Environmental Impact Assessment Methodologies for Use in Life Cycle Assessment (2010) Background Document; 115pp.
- Framework and Requirements for Life Cycle Impact Assessment Models and Indicators (2010) EUR 24586 EN; 116pp.

- Recommendations for Life Cycle Impact Assessment in the European Context – Based on Existing Environmental Impact Assessment Models and Factors (2011) EUR 24571 EN; 159pp.
- Review Schemes for Life Cycle Assessment (2010) EUR 24710 EN; 34pp.
- Reviewer Qualification for Life Cycle Inventory Data Sets (2010) EUR 24379EN; 34pp.

International Standards Organization (ISO) Genève, Switzerland

Development of the international standards for LCA began in the nineties as part of the ISO 14000 family of environmental standards. A revision in 2006 resulted in the core standards ISO 14040 and ISO 14044 that are still valid today. These are accompanied by technical reports.

- ISO 14040 (2006). Environmental Management – LCA – Principles and Framework, International Standard.
- ISO 14044 (2006) Environmental Management – LCA – Requirements and Guidelines, International Standard.
- ISO 14047 (2003) Examples of Application of ISO 14042, Technical Report.
- ISO 14048 (2001) Data Documentation Format, Technical Report.
- ISO 14049 (2000) Examples of Application of ISO 14041 to Goal and Scope Definition and Inventory Analysis, Technical report.

Pré Consultants' LCA Discussion List

http://www.pre-sustainability.com/lca-discussion-list

The LCA discussion list is a global platform to discuss issues related to life cycle assessment and related sustainability issues. On a regular basis, LCA experts and practitioners make important contributions regarding methodology, the sharing of data, and important events in the LCA community. It is also an opportunity to ask questions.

Pré claims the list has over 2500 worldwide members. There is no cost for joining the list.

Society of Environmental Toxicology and Chemistry (SETAC)

SETAC's working groups and workshops have advanced both the application and reputation of LCA by authoring LCA pub;ications, supporting the development of LCA srandardization, partnering with UNEP, and advancing the use of LCA in various sectors.

- A Technical Framework for Life Cycle Assessments (1991) J Fava, R Denison, B Jones, MA Curran, B Vigon, S Sulke, and J Barnum (eds); report from workshop in Smugglers Notch, Vermont, USA, August 18-23, 1990; ISBN 978-1880611005; 152pp.
- Life Cycle Assessment: Inventory, Classification, Valuation, and Data Bases, report from a workshop in Leiden, The Netherlands, December 2-3, 1991,

- Guidelines for Life Cycle Assessment: A Code of Practice (1993) Frank Consoli, David Allen, Ian Boustead, James Fava, William Franklin, Allan Jensen, Nick de Oude, Rod Parrish, Rod Perriman, Dennis Postlethwaite, Beth Quay, Jacinthe Séguin, Bruce Vigon (eds); report from workshop in Sesimbra, Portugal, March 31-April 3, 1993; ISBN 978-90-5607-003-8; 79pp.
- Conceptual Framework for Life-Cycle Impact Assessment (1993) James Fava, Frank Consoli, Richard Denison, Kenneth Dickson, Tim Mohin, Bruce Vigon (eds); report from workshop in Sandestin, Florida, USA, February 1-7, 1992; 188pp.
- Integrating Impact Assessment into LCA (1994) Helias A. Udo de Haes, Allan A. Jensen, Walter Klöpffer, Lars-Gunnar Lindfors (eds); SETAC-Europe, Brussels, Belgium. 198pp.
- Allocation in LCA (1994) Gjalt Huppes and F Schneider; workshop proceedings; SETAC-Europe, Brussels, Belgium.
- Life Cycle Assessment Data Quality: A Conceptual Framework (1994) James Fava, Allan Jensen, Lars Lindfors, Steven Pomper, Bea De Smet, John Warren, Bruce Vigon (eds); report from workshop in Wintergreen, Virginia, USA, October 4-9, 1992; 157pp.
- Towards a Methodology for Life Cycle Impact Assessment (1996) Udo de Haes H.A. (ed) SETAC-Europe, Brussels, Belgium; ISBN 90-5607-005-3; 98pp.
- Simplifying LCA: Just a Cut? (1997) Kim Christiansen (ed); LCA Screening and Streamlining Workgroup; ISBN 978-90-5607-006-9; 53pp.
- Public Policy Applications of Life Cycle Assessment (1997) David Allen, Frank Consoli, Gary Davis, James Fava, John Warren (eds); report from workshop in Wintergreen, Virginia, USA, August 14-19, 1995; 978-1880611180; 127pp.
- Life Cycle Impact Assessment: State-of-the-Art (1998) Larry Barnthouse, James Fava, Ken Humphreys, Robert Hunt, Scott Noesen, Greg Norris, J Willie Owens, Joel A Todd, Bruce Vigon, Keith Weitz, John Young (eds); working group report.
- Streamlined Life-Cycle Assessment: A Final Report from the SETAC North America Streamlined LCA Workgroup (1999) Todd JA and Curran MA (eds); 31pp.
- Code of Life Cycle Inventory Practice (2003) Angeline de Beaufort-Langeveld, Rolf Bretz, Gert van Hoof, Roland Hischier, Pascale Jean, Toini Tanner, Mark Huijbregts (eds); ISBN 1-880611-58-9; 192pp.
- LCA in Buildings and Construction (2003) S Kotaji, A Schuurmans, S Edwards (eds) ISBN 1-880611-59-7.
- Life Cycle Management (2004) David Hunkeler, Konrad Sauer, Gerald Rebitzer, Matthias Finkbeiner, Wulf-Peter Schmidt, Allan Jensen, Heidi Stranddorf, Kim Christiansen (eds) ISBN 9781880611777; 85pp.
- The Working Environment in LCA (2004) P Poulsen, AA Jensen, A-B Antonsson, G Bengtsson, M Karling, A Schmidt, O Brekke, J Brecker, AH Verschoor (eds) ISBN 1-880611-68-6.

- Scenarios in LCA (2004) Gerald Rebitzer and Thomas Ekvall (eds) ISBN 1-886011-57-0; 88pp.
- Global Guidance Principles for LCA Databases: A Basis for Greener Processes and Products (2011); Guido Sonnemann and Bruce Vigon (eds); report from workshop in Shonan Village, Japan; 160pp.

UNEP/SETAC Life Cycle Initiative

The Life Cycle Initiative aims to foster the concept of Life Cycle Thinking (LCT). Learn more about the idea behind LCT and existing methodologies. http://www.lifecycleinitiative.org

Publications:

- Life Cycle Thinking in Latin America (2015) English; 16pp.
- An Analysis of Life Cycle Assessment in Packaging for Food and Beverage Applications (2013) English; 81pp.
- The Methodological Sheets for Sub-Categories in Social Life Cycle Assessment (S-LCA) (2013) English; 144pp.
- Greening the Economy through Life Cycle Thinking (2012) English; 60pp.
- Towards a Life Cycle Sustainability Assessment (2011) English; 86pp.
- Global Guidance Principles for LCA Databases: A Basis for Greener Processes and Products (2011) English; 160pp.
- Guidelines for Social LCA of Products (2009) English, French and Dutch versions; 104pp.
- LCM: How Business uses it to Decrease Footprint, Create Opportunities and Make Value Chains More Sustainable (2009) English; 48pp.
- Guidance on How to Move from Current Practice to Recommended Practice in Life Cycle Impact Assessment (2008) English; 33pp.
- Communication of Life Cycle Information in the Building and Energy Sectors (2008) English; 93pp.
- Life Cycle Management: a Business Guide to Sustainability (2007) English; 52pp.
- Background Report for a UNEP Guide to Life Cycle Management (2006) English; 108pp.
- Life Cycle Approaches: the Road from Analysis to Practice (2005) English; 89pp.
- Why Take a Life Cycle Approach (2004) Chinese, English, Spanish, French and Japanese versions; 28pp.
- Report of the LCM Definition Study Peer Review (2003) English; 79pp.
- Life Cycle Impact Assessment Methods (2003) English; 25pp.
- Evaluation of Environmental Impacts in Life Cycle Assessment (2003) English; 108pp.

LCA Centers and Societies[1]

African LCA Network (ALCANET)

http://www.estis.net/sites/alcanet/

ALCANET is a regional network aiming to support the research process, to provide teaching and training activities, and to foster public understanding of Life Cycle Assessment (LCA) in Africa. ALCANET may no longer be active.

American Center for LCA (ACLCA)

http://lcacenter.org/

ACLCA is a non-profit membership organization that seeks to build capacity and knowledge of Environmental LCA among industry, government and NGOs. ACLCA manage and administer a certification program for LCA Practitioners. Until 2015, ACLCA held the annual LCA conference series in North America.

Asia LCA Agri-Food Network

http://www.lcaagrifoodasia.org/Network.html

Potential applications of LCA in the agri-food sector have been recognized internationally, while LCA is rather new in the Asian region. As a result, the initiative on "LCA Agri-Food Asia" is being developed with the multi-national collaboration from governmental organizations, higher research and educational institutes, as well as private companies.

Association for Life Cycle Assessment in Latin America (ALCALA) Asociación Latino Americana para la Evaluación del Ciclo de Vida

http://alcalacr.org (in Spanish)

Australian Life Cycle Assessment (ALCAS)

www.alcas.asn.au

ALCAS was established in 2001 to promote life cycle practices and sustainable development, and to coordinate the rapidly growing professional community in Australia. A not-for-profit organization, ALCAS has individual and corporate members from industry, government, academia and service organizations. They welcome membership from people interested in the practice, use, development, education, interpretation of and advocacy for life cycle based approaches.

China Life Cycle Initiative (CNLCI)

http://www.cnlci.net/

The CNLCI is a major initiative by Ecovane and its partners. The aim is to position China research and activities at the leading edge of LCA and LCA theory and practice, through education and providing and maintaining a national, publicly-accessible database with easy access to authoritative,

[1] For a more comprehensive listing, refer to "Mapping and Characterization of LCA Networks" (2012) Anders Bjørn, Mikolaj Owsianiak, Alexis Laurent, Christine Molin, Torbjørn Bochsen Westh, Michael Zwicky Hauschild, Int J Life Cycle Assess 18:812-827; doi:10.1007/s11367-012-0524-6.

comprehensive and transparent environmental information on a wide range of Chinese products and services over their life cycle.

Red Colombiana de Análisis de Ciclo de Vida (Columbian Network of LCA)
http://www.redacvcolombia.blogspot.com/

LCA Center Denmark
http://www.lca-center.dk

[avnir] LCA Platform
http://www.avnir.org

The [avniR] LCA Platform helps organizations develop better products, systems and services through life cycle management. They provide direct support for organizations in Northern France, and develop partnerships with national and international organisations to share tools and create new projects. [avniR] is an initiative of the not-for-profit organisation cd2e team of environmental and business experts dedicated to supporting the environmental sector in Northern France and abroad. The [avniR] Platform organizes the annual LCA [avniR] Conference.

LCA Center Hungary
http://www.lcacenter.hu/

Indian Society for LCA (ISLCA)
http://neef.in/islca.html

- Capacity building for development of LCA in India through its courses, training programmes, conferences, seminars, research projects etc.
- Integrating socio-economic concepts in LCA.
- Representing India in national and international forums on LCA and related areas.
- Networking with leading professionals in LCA and related fields.
- Promoting publications of the ISLCA including its planned periodicals and newsletters, publications, videotapes, discs and other communication media.

Red Iberoamericana de Ciclo de Vida (Ibero-American Network oLCA)
https://rediberoamericanadeciclodevida.wordpress.com/ (in Spanish)

Israel LCA Network (LinkedIn)
https://www.linkedin.com/groups/Israel-LCA-network-5073827

Italian LCA Network (Rete Italiana LCA)
http://www.reteitalianalca.it/

LCA Society of Japan (JLCA)
http://lca-forum.org/english/

Established in October 1995 as a network to distribute information on LCA to interested parties in industry, academia and government, JLCA has promoted discussions on the future progress of LCA in Japan, and released a JLCA report and LCA Position Statement in June 1997 proposing the construction of an LCA background database as well as the development of impact assessment methods that are most suitable for the circumstances in Japan. In 1998, the Ministry of Economy, Trade and Industry began a 5 year LCA Project, which was completed at the end of March 2002 with the construction of an LCA database. In consideration of the circumstances of the project and the provision of data for inclusion in the database from more than 50 industry organizations, the Ministry of Economy, Trade and Industry decided that it would be desirable for this database to be managed and operated by the JLCA , since the group brings together industry, academia and government.

South and South-East Asia (SEASIA) Network of the Life Cycle Initiative

http://www.estis.net/sites/seasia/

SEASIA Network represents UNEP-SETAC's Life Cycle Initiative programme in South and South-East Asian Countries.

Korean Society for LCA (KSLCA)

http://kslca.com/ (in Korean)

Along with the dynamic activities of government, industry, and academia with respect to LCA, the Korean Society for Life Cycle Assessment (KSLCA) was established in 1997 aiming to be the conduit for distributing LCA to different sectors of society and for reporting its research activities, contributing significantly to the strengthening of the Korean LCA infrastructure. This group may no longer be active.

Koran Society for Industrial Ecology

http://www.ksie.re.kr/ (in Korean)

LCA Malaysia

http://lcamalaysia.sirim.my/

LCA Malaysia was created in 2008, in connection with the National LCA Project for the establishment of the national LCA database. This effort was mandated to SIRIM Berhad by the Government of Malaysia under the Ninth Malaysia Plan (2005-2010) to support cleaner production approach in the country's industrial activities.

Red Mexicana de Análisis de Cyclo de Vida (Mexican Network for LCA)

http://sitios.iingen.unam.mx/CicloDeVida/Default.htm (in Spanish)

Life Cycle Association of New Zealand (LCANZ)

http://lcanz.org.nz/links/lca

The Life Cycle Association of New Zealand (LCANZ) was established in June 2009 to provide a focal point for Life Cycle Assessment and Management work conducted in New Zealand. It aims to promote networking and knowledge sharing between organisations and people. It will also promote and raise the awareness of Life Cycle Thinking among the wider public and business.

Nordic Life Cycle Association (NorLCA)

http://www.norlca.org

NorLCA is a Nordic association of individuals, companies and organizations working with life cycle approaches broadly defined. This covers life cycle thinking, life cycle design, life cycle management, life cycle costs, life cycle assessment, product oriented management, Integrated Product Policy - IPP, green procurement, green marketing etc.

Red Peruana Ciclo de Vida (Peruvian LCA Network)

http://www.red.pucp.edu.pe/ciclodevida/

Swedish Life Cycle Center

http://lifecyclecenter.se/

The Swedish Life Cycle Center (formerly CPM) is a center of excellence for the advancement of life cycle thinking in industry and other parts of society. It has been in existence since 1996. The Center works with research, implementation, communication and exchange of experience on life cycle management. Characterized by close and continuous interaction between academia, applied research institutes, industry and authorities, the Center's mission is to improve the environmental performance of products and services, as a natural part of sustainable development. Partners include industries and research groups with profound commitment to sustainable development and life cycle thinking. Also the need for an open exchange of information and knowledge between organizations remains.

Swiss Discussion Forum on LCA

http://www.lcaforum.ch/

Thai LCA Network

http://thailca.net/

Facebook page: https://www.facebook.com/ThLCA?filter=1

Glossary

Allocation – see Co-Product Allocation

Attributional LCA - System modelling approach in which inputs and outputs are attributed to the functional unit of a product system by linking and/or partitioning the unit processes of the system according to a normative rule.

Background - The data or part of the LCA model outside immediate control of the decision maker (see Foreground).

Biogenic carbon - Carbon that is removed from the atmosphere and incorporated into the physical mass of a plant or organism.

Characterization –The calculation of category indicator results with characterization factors.

Characterization Factor – A multiplier derived from models which are then used to convert life cycle inventory to the common unit of the impact category, thus allowing calculation of a category indicator result.

Classification – The assignment of life cycle inventory results to impact categories.

Cleaner Production - The continuous application of an integrated preventative environmental strategies for processes, products and services to increase efficiency and reduce risks to humans and the environment (see also, Pollution Prevention).

Closed Loop recycling – When a product is recycled (postconsumer) to make more of the same product.

Comparative Assertion – Making an environmental claim regarding the superiority of one product over another in a public forum.

Consequential LCA - System modelling approach in which activities in a product system are linked so that activities are included in the product system to the extent that they are expected to change as a consequence of a change in demand for the functional unit.

Contribution Analysis – The determination of which inventory data or impact indicator has the biggest influence on the final results by comparing the contribution of the life cycle stages or groups of processes to the total, and examining them for relevance.

Co-Product Allocation – Assignment of inputs and outputs (life cycle inventory) to any of two or more products coming from the same unit process or product system.

Cradle-to-Cradle – Study boundaries begin with raw material production and extend through manufacturing, on to end-of-life management, including any reuse or recycling that may occur.

Cradle-to-Gate – Study boundaries begin with raw material production and extend to the point where the product leaves a (manufacturing) facility.

Critical Review - Process intended to ensure consistency between a life cycle assessment and the principles and requirements of the International Standards on Life Cycle Assessment.

Cut-Off Criteria – The omission of negligible inputs or activities within the system being studied with the intent of simplification.

Economic Input/Output (EIO) Analysis – An economic discipline that models the interdependencies of production and consumption between industries and households within a nation's economy.

Foreground – The data or part of the LCA model directly under the control of the decision maker (see Background).

Function – The purpose the system under study fulfills.

Functional Unit – The amount of material or number of items needed to meet the system function. It provides a reference to which the inputs and outputs are related.

Gate-to-Gate – Study boundaries are drawn around a single operation, typically a facility or factory.

Goal and Scope - The first phase of an LCA; establishing the aim of the intended study, the functional unit, the reference flow, the product system(s) under study and the breadth and depth of the study in relation to this aim.

Grouping - The assignment of impact categories into one or more sets.

Interpretation - The phase of LCA in which the findings of the inventory analysis and impact assessment are evaluated with relation to the goal of the study in order to reach conclusions and, if desired, make recommendations.

Life Cycle Assessment - The compilation and evaluation of the inputs, outputs and potential environmental impacts of a product system throughout its life cycle.

Life Cycle Impact Assessment - The phase of LCA aimed at understanding and evaluating the magnitude and significance of the potential environmental impacts related to a product system throughout its life cycle.

Life Cycle Inventory Analysis - The phase of LCA involving the compilation and quantification of inputs and outputs for a product throughout its life cycle.

Life Cycle Sustainability Assessment - the evaluation of all environmental, social and economic negative impacts and benefits in decision-making processes towards more sustainable products throughout their life cycle.

Normalization - The calculation of the magnitude of category indicator results relative to a reference. It is intended to provide context and relative magnitude of an impact indicator.

Open Loop Recycling - When a product is recycled (postconsumer) into a different product that the one it was originally.

Pollution Prevention - Systematic, periodic internal reviews of specific processes and operations designed to identify and provide information about opportunities to reduce the use, production, and generation of toxic and hazardous materials and waste. This term is used more commonly in the US for Cleaner Production.

Reference Flow - A measure of the outputs from processes in a product system required to fulfill the functional unit.

Resource and Environmental Profile Analysis (REPA) - An early term preceding LCA, coined by MRI, for a system analysis of the investigated product from cradle to grave.

Sensitivity Analysis -Systematically changing an input parameter and observing the impact on the results as a way to verify the effects of models and assumptions on the outcome of the study.

Social Life Cycle Assessment - The compilation and evaluation, of the inputs, outputs and potential social impacts in relation to five main stakeholder categories (workers, local community, society, consumers, value chain actors) of a product system throughout its life cycle.

Supply Chain - The connection of organizations involved in the transformation of natural resources and components into a finished product intended for delivery an end customer. In sophisticated supply chain systems, used products may re-enter the supply chain at any point where residual value is recyclable.

System Expansion - An approach to dealing with co-product allocation by including an additional product function equivalent to a co-product's. Also used to give credit for a product or material that is displaced by the product or material of interest.

Transparency - The open, comprehensive and understandable presentation of information.

Uncertainty Analysis - A systematic procedure to quantify the uncertainty introduced in the results of a life cycle inventory analysis due to the cumulative effects of model imprecision, input uncertainty and data variability.

Unit Process - The smallest element modeled in a life cycle inventory analysis for which inputs and output data are quantified.

Waste Minimization - Source reduction and/or environmentally sound recycling methods prior to energy recovery, treatment, or disposal of wastes, but excluding waste treatment, such as compacting, neutralizing, diluting, and incineration.

Weighting – The application of value judgments, such as social and political priorities, to the characterization indicator results or to their normalized version.

Also of Interest

By the Same Author

Life Cycle Assessment Handbook: A Guide for Environmentally Sustainable Products, edited by Mary Ann Curran, ISBN 9781118099728. The first book of its kind, the LCA Handbook will become an invaluable resource for environmentally progressive manufacturers and suppliers, product and process designers, executives and managers, and government officials who want to learn about this essential component of environmental sustainability. *NOW AVAILABLE!*

Check out these other titles from Scrivener Publishing

i-Smooth Analysis: Theory and Applications, by A.V. Kim, ISBN 9781118998366. A totally new direction in mathematics, this revolutionary new study introduces a new class of invariant derivatives of functions and establishes relations with other derivatives, such as the Sobolev generalized derivative and the generalized derivative of the distribution theory. *DUE OUT IN MAY 2015.*

Reverse Osmosis: Design, Processes, and Applications for Engineers 2nd Edition, by Jane Kucera, ISBN 9781118639740. This is the most comprehensive and up-to-date coverage of the "green" process of reverse osmosis in industrial applications, completely updated in this new edition to cover all of the processes and equipment necessary to design, operate, and troubleshoot reverse osmosis systems. *DUE OUT IN MAY 2015.*

Pavement Asset Management, by Ralph Haas and W. Ronald Hudson, with Lynne Cowe Falls, ISBN 9781119038702. Written by the founders of the subject, this is the single must-have volume ever published on pavement asset management. *DUE OUT IN MAY 2015.*

Open Ended Problems: A Future Chemical Engineering Approach, by J. Patrick Abulencia and Louis Theodore, ISBN 9781118946046. Although the primary market is chemical engineers, the book covers all engineering areas so those from all disciplines will find this book useful. *NOW AVAILABLE!*

Fracking, by Michael Holloway and Oliver Rudd, ISBN 9781118496329. This book explores the history, techniques, and materials used in the practice of induced hydraulic fracturing, one of today's hottest topics, for the production of natural gas, while examining the environmental and economic impact. *NOW AVAILABLE!*

Formation Testing: Pressure Transient and Formation Analysis, by Wilson C. Chin, Yanmin Zhou, Yongren Feng, Qiang Yu, and Lixin Zhao, ISBN 9781118831137. This is the only book available to the reservoir or petroleum engineer covering formation testing algorithms for wireline and LWD reservoir analysis that are developed for transient pressure, contamination modeling, permeability, and pore pressure prediction. *NOW AVAILABLE!*

Electromagnetic Well Logging, by Wilson C. Chin, ISBN 9781118831038. Mathematically rigorous, computationally fast, and easy to use, this new approach to electromagnetic well logging does not bear the limitations of existing methods and gives the reservoir engineer a new dimension to MWD/LWD interpretation and tool design. *NOW AVAILABLE!*

Desalination: Water From Water, by Jane Kucera, ISBN 9781118208526. This is the most comprehensive and up-to-date coverage of the “green” process of desalination in industrial and municipal applications, covering all of the processes and equipment necessary to design, operate, and troubleshoot desalination systems. *NOW AVAILABLE!*

Tidal Power: Harnessing Energy From Water Currents, by Victor Lyatkher, ISBN 978111720912. Offers a unique and highly technical approach to tidal power and how it can be harnessed efficiently and cost-effectively, with less impact on the environment than traditional power plants. *NOW AVAILABLE!*

Electrochemical Water Processing, by Ralph Zito, ISBN 9781118098714. Two of the most important issues facing society today and in the future will be the global water supply and energy production. This book addresses both of these important issues with the idea that non-usable water can be purified through the use of electrical energy, instead of chemical or mechanical methods. *NOW AVAILABLE!*

Biofuels Production, Edited by Vikash Babu, Ashish Thapliyal, and Girijesh Kumar Patel, ISBN 9781118634509. The most comprehensive and up-to-date treatment of all the possible aspects for biofuels production from biomass or waste material available. *NOW AVAILABLE!*

Biogas Production, Edited by Ackmez Mudhoo, ISBN 9781118062852. This volume covers the most cutting-edge pretreatment processes being used and studied today for the production of biogas during anaerobic digestion processes using different feedstocks, in the most efficient and economical methods possible. *NOW AVAILABLE!*

Bioremediation and Sustainability: Research and Applications, Edited by Romeela Mohee and Ackmez Mudhoo, ISBN 9781118062845. Bioremediation and Sustainability is an up-to-date and comprehensive treatment of research and applications for some of the most important low-cost, “green,” emerging technologies in chemical and environmental engineering. *NOW AVAILABLE!*

Sustainable Energy Pricing, by Gary Zatzman, ISBN 9780470901632. In this controversial new volume, the author explores a new science of energy pricing and how it can be done in a way that is sustainable for the world’s economy and environment. *NOW AVAILABLE!*

Green Chemistry and Environmental Remediation, Edited by Rashmi Sanghi and Vandana Singh, ISBN 9780470943083. Presents high quality research papers as well as in depth review articles on the new emerging green face of multidimensional environmental chemistry. *NOW AVAILABLE!*

Energy Storage: A New Approach, by Ralph Zito, ISBN 9780470625910. Exploring the potential of reversible concentrations cells, the author of this groundbreaking volume reveals new technologies to solve the global crisis of energy storage. *NOW AVAILABLE!*

Bioremediation of Petroleum and Petroleum Products, by James Speight and Karuna Arjoon, ISBN 9780470938492. With petroleum-related spills, explosions, and health issues in the headlines almost every day, the issue of remediation of petroleum and petroleum products is taking on increasing importance, for the survival of our environment, our planet, and our future. This book is the first of its kind to explore this difficult issue from an engineering and scientific point of view and offer solutions and reasonable courses of action. *NOW AVAILABLE!*